A-14 Sprite 'Mk IV' - Red - P.74, 'Tartan Red' with black trim.
A-14 Sprite Mk IV Blue - P.86
HAN 9/10, which? HAN 9, 1966-1969.(ozr).

ORIGINAL
SPRITE & MIDGET

Other titles available in the *Original* series are:

Original AC Ace & Cobra
by Rinsey Mills
Original Aston Martin DB4/5/6
by Robert Edwards
Original Austin Seven
by Rinsey Mills
Original Austin-Healey (100 & 3000)
by Anders Ditlev Clausager
Original Citroën DS
by John Reynolds with Jan de Lange
Original Corvette 1953-1962
by Tom Falconer
Original Ferrari V8
by Keith Bluemel
Original Honda CB750
by John Wyatt
Original Jaguar XK
by Philip Porter
Original Jaguar E-Type
by Philip Porter
Original Jaguar Mark I/II
by Nigel Thorley
Original Land-Rover Series I
by James Taylor
Original Mercedes SL
by Laurence Meredith
Original MG T Series
by Anders Ditlev Clausager
Original MGA
by Anders Ditlev Clausager
Original MGB
by Anders Ditlev Clausager
Original Mini Cooper and Cooper S
by John Parnell
Original Morgan
by John Worrall and Liz Turner
Original Morris Minor
by Ray Newell
Original Porsche 356
by Laurence Meredith
Original Porsche 911
by Peter Morgan
Original Porsche 924/944/968
by Peter Morgan
Original Triumph TR2/3/3A
by Bill Piggott
Original Triumph TR4/4A/5/6
by Bill Piggott
Original Vincent
by J. Bickerstaff
Original VW Beetle
by Laurence Meredith
Original VW Bus
by Laurence Meredith

ORIGINAL
SPRITE & MIDGET

Terry Horler

Photography by John Colley
Edited by Mark Hughes

Bay View Books

FRONT COVER
Early and late versions from the Sprite part of the story, with David Whyley's Leaf Green MkI playing second fiddle to Craig Polly's New Racing Green MkIV. Minor departures from pure originality on the MkIV are the addition of sun visors (for practical reasons), extra motifs on the wheel centres, and the absence of the small British Leyland badge which would normally be fitted to the lower rear corner of the front wing.

BACK COVER
Jonathan Whitehouse-Bird's Tartan Red Midget MkII, at rest in a period garage forecourt, is original and unrestored.

Published 1994 by Bay View Books Ltd
The Red House, 25-26 Bridgeland Street
Bideford, Devon EX39 2PZ

Reprinted 1996
Reprinted 1998

© Copyright 1994 by Bay View Books Ltd

All rights reserved. No part of this publication may be reproduced or transmitted in any form or by any means, electronic or mechanical, including photocopying, recording or in any information storage or retrieval system, without the prior written permission of the publisher.

Designed by Nick Kisch
Typesetting and layout by Chris Fayers

ISBN 1 870979 45 1
Printed in Hong Kong
by Paramount Printing Group

CONTENTS

Introduction	6
Sprite and Midget Past and Present	8
Sprite MkI (1958-61)	**16**
Identification, Dating and Production	41
Options, Extras and Accessories	43
Colour Schemes	47
Production Changes	47
Sprite MkII and Midget MkI (1961-64)	**50**
Identification, Dating and Production	66
Options, Extras and Accessories	67
Colour Schemes	69
Production Changes	70
Sprite MkIII and Midget MkII (1964-66)	**74**
Identification, Dating and Production	83
Options, Extras and Accessories	84
Colour Schemes	85
Production Changes	85
Sprite MkIV and Midget MkIII (1966-74)	**86**
Identification, Dating and Production	110
Options, Extras and Accessories	113
Colour Schemes	115
Production Changes	115
Midget 1500 (1974-79)	**118**
Identification, Dating and Production	134
Options, Extras and Accessories	135
Production Changes	135
Colour Schemes	137
Innocenti versions (1960-70)	**138**
Buying and Restoration	**143**
Clubs	144

INTRODUCTION

In September 1969 I bought my first Austin-Healey Sprite, a 1959 MkI in Old English White. It was not the fulfilment of a life-long ambition to own such a car – I needed transport and it was cheap! I soon found out why it was cheap. Ten years and as many owners had all left their mark, leaving some poor, unfortunate, subsequent owner – me! – to sort out all the problems. I learned a lot about that car very quickly, but I also found it great fun to drive.

Quite why I sold it only eight months later remains one of those little mysteries in life. The lure of something faster or selling it for more than I paid may have been good enough reasons at the time, but as time passed I regretted it. Realising my mistake, my next Sprite, a 1960 model, arrived in 1976. This was an even bigger mistake than my decision to sell my first Sprite: I didn't realise when I handed over the cash, but the bodyshell was twisted following some previous owner's misadventure. My third Sprite, another 1960 model, was acquired in 1979 and this is the one which I still own to this day. It has been joined by a 1968 MkIV Sprite which provides more practical, if not so characterful, everyday transport.

In case you're thinking I've ignored MG-badged cars, a brief period of ownership of a 1966 MkII Midget provided a summer of wind-in-the-hair fun, but like so many other Sprites and Midgets in the early 1980s it didn't survive the winter. For every Sprite or Midget that I have owned and used, I must also have dismantled three others to keep them going, a fact that perhaps I shouldn't be too proud of today!

Having established that these cars held a special fascination for me, and knowing that I wasn't alone in being attracted to them, the idea arose with a small group of like-minded owners to form a club exclusively for Sprites and Midgets. I have had the honour, they tell me, of being the Midget & Sprite Club's General Secretary and Archivist since the beginning, in 1983. Over the years, this has brought me into contact with hundreds of owners and their cars, and a vast amount of information has been gathered.

Many people know the basics about these cars, that there were 'Frogeye' Sprites, 'chrome-bumpered' Spridgets and 'rubber-bumpered' 1500 Midgets. More precisely, however, there are 14 distinct versions of Sprites and Midgets, plus export variations. In accepting the challenge to write this book, I set out to describe in detail each of the various versions and how they developed over 21 years of production.

This book would not have been possible without the knowledge and help of a number of people to whom I give credit and my thanks for their hours of toil and patience. I would like especially to thank Jonathan Whitehouse-Bird, Rudolf Probst and Peter Seamen for their help with the pre-1964 versions, Ray English and Tony Slattery for the Australian content, and Bob Zimmerman, Tom Christensen and Gil Gibbons for the North American information and photographs. I must also thank Stuart Granger and Dave Gilbert for helping with research and hours of sifting through the factory records at the British Motor Industry Heritage Trust.

The BMIHT has kindly allowed reproduction of the information gathered. Special thanks to its Archivist, Anders Clausager, who has not only been most helpful and accommodating on my visits, but has also contributed his own research and knowledge to this book. My two typists, Sylvia Horler and Pam Higgs, successfully transcribed my appalling handwritten notes into something readable for my editor, Mark Hughes. I cannot get away without thanking the Midget & Sprite Club, for without its existence I would not have been able to tackle such a task.

Finally, a big 'thank you' to the owners of the cars

Original literature is a good finishing touch to any car.

This 'Frogeye' with rarely-seen rubber floor matting is typical of the high standard of originality portrayed in this book.

INTRODUCTION

Production of this book required a painstaking search for cars in highly original condition. The mileage of 7675 recorded on Allan Cameron's beautifully preserved 'Frogeye' is genuine.

One of the most surprising cars featured in this book is the 1962 Sprite MkII owned since new by Roy de St Croix. Its startling Deep Pink paint, a factory colour, was chosen by so few customers that this surviving car may well be unique.

used for photography and expertly captured on film by John Colley. Their names and their cars are listed in the adjacent panel. John's work was supplemented by a few additional photographs taken by Ron Middleton of Focalpoint, David Whyley, John Bowman, Peter Seamen, Colin Jarvis, Steve Lloyd and Dennis Acton. Just when I was despairing of being able to locate a Sprite MkIII good enough to provide some elusive detail views, Malcolm McKay rummaged through the files at *Classic Cars* magazine

and found a handful of suitable shots.

Following established practice with the *Original* series, a viewing session was held so that all the photographs in this book could be vetted by a panel of experts. Five of those already mentioned – Anders Clausager, Jonathan Whitehouse-Bird, Peter Seamen, Stuart Granger and Dave Gilbert – joined me in this exercise, together with John Mead, who has immense experience as a restorer of Sprites and Midgets and as a concours judge for the Austin-Healey Club. John also kindly read the manuscript.

Throughout the descriptive chapters, I refer to the various versions by their chassis prefix letters and numbers. This is both the simplest and most accurate means of identification rather than becoming misled by the 'Mk' numbers which, in certain cases, do not clearly define the model without further descriptive additions. Yes, I can live with the descriptions 'Frogeye' Sprites and 'rubber-bumpered' Midgets, but 'chrome-bumpered' Spridgets is rather akin to saying that a television works by electricity – true but just a little short in detail. In referring to the left- and right-hand sides in the following chapters, this always applies to viewing the car from the rear.

I should cover myself by saying that it has not been possible to contain every single, fact, figure, nut, bolt and grease nipple in these chapters. Certain items have been omitted by design, others because they have escaped me. I have endeavoured, however, not to include inaccuracies. For the most part, the information has come from manufacturer's publications, contemporary information and from reliable experts. Where I may have misfired in print, I would be grateful to receive comments via the publisher.

Terry Horler
Yate, Bristol
November 1993

OWNERS OF FEATURED CARS

Model	Colour	Registration	Owner
Sprite MkI (AN5)	Old English White	964 FRT	Allan Cameron
Sprite MkI (AN5)	Old English White	ODY 128	Bob Reid
Sprite MkI (AN5)	Primrose Yellow	WGJ 837	Ian Wilson
Sprite MkI (AN5)	Iris Blue	5983 AC	Peter Seamen
Sprite MkI (AN5)	Leaf Green	1918 K	David Whyley
Sprite MkII (HAN6)	Deep Pink	XDR 23	Roy de St Croix
Sprite MkIII (HAN8)	Tartan Red	MGT 401D	Barry Knight
Sprite MkIII (HAN8)	Tartan Red	DNF 293C	John Dunn
Sprite MkIII (HAN8)	Riviera Blue	ACC 496B	Phil Lynford
Sprite MkIV (HAN 9)	Mineral Blue	MMJ 626H	Clive Wagstaff
Sprite MkIV (HAN10)	Green	LXC 406H	Craig Polly
Sprite MkIV (HAN10)	Old English White	AJW 674H	Stuart Granger
Austin Sprite (AAN10)	New Racing Green	RUU 260J	Dee Houghton
Midget MkI (GAN1)	Old English White	6846 PE	Jonathan Whitehouse-Bird
Midget MkI (GAN1)	Old English White	9961 PT	Joy Dowling
Midget MkI (GAN2)	Old English White	DMU 9A	Andy Ricketts
Midget MkI (GAN2)	Tartan Red	JLJ 434E	Bruce Halliday
Midget MkII (GAN3)	Tartan Red	101 XKX	Jonathan Whitehouse-Bird
Midget MkIII (GAN4)	Old English White	NBD 320F	Dave Gilbert
Midget MkIII (GAN5)	Mallard Green	RDM 584M	Denis Matthews
Midget MkIII (GAN5)	Flame Red	586 UMX	Bob Zimmerman
Midget 1500 (GAN6)	Jubilee Edition	OBW 50P	Michael Cohen
Midget 1500 (GAN6)	Black	943 HBC	Bob Barker
Midget 1500 (GAN6)	Brooklands Green	C MY MGGO	Tom Christensen
Innocenti	Red	GNB 902D	Neil Roger

PAST AND PRESENT

On 20 May 1958, in the sunshine of Monte Carlo, members of the press were introduced to a new small sports car – the Austin-Healey Sprite. 'Cheap and cheerful' in the very best sense of the phrase, the Sprite was an unbeatable package which offered more fun for the money than any car of the time – and arguably ever since. Rave reviews showed how the press loved it, and so too did a new generation of sports car enthusiasts.

One rather unusual feature of the clean and simple styling was to earn the new car an instant nickname. Although the Sprite was originally intended to have pop-up headlamps, production expediency and cost led this feature to be replaced by headlamps in fixed pods rising from the top of the bonnet. Hardly surprising, then, that 'Frogeye' – or 'Bugeye' in the USA – became the Sprite's unofficial but almost universal nickname. Complementing the 'Frogeyes', the radiator grille joined in the fun by simulating a broad-grinning mouth. Cartoonists of the day were quick to exploit this, portraying the Sprite as a friendly and mischievous little character just asking to be enjoyed. In fact, the Sprite had character in abundance.

The success was not to be short-lived, for the Sprite began a production series that was to last 21 years. Although many changes occurred during this period, with the original 'Frogeye' form disappearing in 1961, the basic concept and structural building blocks remained the same throughout.

The Sprite was the product of a partnership between Austin, which in 1952 merged with the Nuffield Organisation to create the British Motor Corporation (BMC), and the Donald Healey Motor Co Ltd. In 1946, with the world looking to a bright new future, Donald Healey had formed his company at the Cape Works in Warwick. His first two models, the Elliott saloon and Westland roadster, both with Riley 2.4-litre engines, offered exceptional performance for the day. Later came the Silverstone, his first open two-seater sports car aimed very much at the

Bob Reid's original and unrestored Sprite MkI in Old English White. Little explanation should be required to understand how the nickname of 'Frogeye' (or 'Bugeye') Sprite came to be applied.

Sprite & Midget Past and Present

'Pinky'. Yes, Deep Pink was a standard colour option. Roy de St Croix has owned this Sprite MkII from new. It was manufactured as a basic model but converted by the supplying dealer to de luxe specification with the addition of bumpers, heater, windscreen washers and an adjustable passenger seat.

competitive driver. A more refined sports car followed in the shape of the Nash-Healey with a straight-six American Nash engine of 3.8 litres, but this did not provide the ideal choice of powerplant. Looking for a more suitable and readily available engine, Healey found a possible solution at Austin.

Austin had designed the A90 Atlantic to capture important export sales in North America. It was a rather garish styling exercise by British standards, but unfortunately for Austin the Americans didn't seem to like it very much either! As a result, there was a ready supply of A90 engines and running gear which Healey could use most effectively in a new sports car shown at the 1952 Earls Court Motor Show.

The new Healey 100 stole the show and aroused the interest of Austin chief Leonard Lord. A deal struck then and there arranged for the Healey Motor Co to act as design and development consultants while Austin would produce the new sports car as the Austin-Healey 100. Being a small company with limited capacity, Healey could not hope to manufacture anything like the quantity which Lord was envisaging, so the Austin factory at Longbridge undertook initial production manufacturing responsibilities. Healey would receive royalties on sales and also undertake a competition programme to further both development and 'image'.

With the partnership formed between Austin and Healey, the next significant move came in early 1956 at a meeting between Lord and Healey. With MG as a part of the BMC empire, the company as a whole was fielding two sports cars on the market – the 2639cc Austin-Healey 100-Six and the smaller 1489cc MGA. Lord felt there was a noticeable gap for a small-engined sports car. "What we need is a bug," he suggested, giving Healey a green light to go away and design a suitable car. As with the 100 model, readily-available components from BMC would be

Andy Ricketts' MG Midget MkI in GAN2 form, with a 1098cc engine, disc front brakes and improved interior trim, finished in Old English White with Hazelnut trim and hood. This trim colour and optional wire wheels were not available on the 948cc Midget MkI. Speedwell exterior door handles are a period accessory.

9

Original Sprite & Midget

an essential requirement, so the 948cc Austin A35 was selected as the main provider.

With Donald's eldest son, Geoffrey, as chief project engineer, Barry Bilbie as chassis designer and Gerry Coker as body designer, the new sports car rapidly began to take shape. Fairly early on, it was realised that some A35 components would not be suitable in their original form, changes to the rear brakes, rear suspension, steering and carburation being required.

The A35's unusual arrangement of rear brake operation, whereby a single chassis-mounted slave cylinder operated both rear brakes through a mechanical linkage, was discarded in favour of a more conventional set-up with a single slave cylinder mounted within each rear brake. The changes to the rear suspension were necessitated by the desire to keep the structural work to an absolute minimum. Instead of a separate chassis frame, the chassis and body were to be one monocoque unit, for the first time in a Healey. To keep weight and production costs down, no suspension attachment points were provided behind the rear axle line. In order to fit the A35 rear axle, semi-elliptic springs were replaced by quarter-elliptic springs which located into the rear edge of the floorpan and rear bulkhead. A radius arm above the axle provided the lateral location required on each side, the forward ends of these arms pivoting in brackets bolted to the rear bulkhead, and the axle requiring suitable brackets to be welded on to accommodate the new springs and arms. The A35's steering box with drag links could not be accommodated and would not have provided particularly precise steering, so this was replaced by the far superior rack and pinion system from the Morris Minor. This feature, giving highly direct and accurate steering without incurring heavy effort, was to be one of the Sprite's most praised attributes. Whereas the A35's 948cc A-Series engine produced 34bhp on a single downdraught Zenith carburettor, the low bonnet line of the Sprite necessitated a change to sidedraught carburation, the answer arriving in a pair of SU H1 1⅛in carburettors which helped to raise output to 42.5bhp at 5200rpm.

Simplicity being a keynote in design, the combined body/chassis unit incorporated some surprising features. There was no opening boot lid in the rear shroud, so the only access to the boot area – which was actually quite generous – was by pivoting the seat backrests forward and diving in through the rear of the cockpit. The doors contained large pockets as no winding windows were provided, and neither external door handles nor a method of locking were provided. The bonnet, although hinged conventionally at its rear edge, lifted complete with the front wings and front panel, leaving the headlights to gaze skywards while under-bonnet matters were attended to. Although this provided good access, the ever-present worry of this heavy steel structure falling shut to imitate a 'man-eating frog' was nothing to smile about, nor was the likelihood of clouting your head on part of the structure.

Phil Lynford's Riviera Blue Sprite MkIII displays a simpler exterior than the parallel Midget, with no waistline or bonnet moulding strips and a simple radiator grille – but the AH-embossed hub caps are more distinctive. New doors with winding windows, quarterlights, exterior door handles and locks were the most obvious additions for the Sprite MkIII and Midget MkII.

In order to use the name Sprite on the new car, permission had to be sought from Daimler, who had acquired the name from Riley. Fortunately, Daimler was no longer planning to use the name and allowed its transfer to the new Austin-Healey. As with its 100 model, Healey would receive royalties on sales and its arrangement to continue development work would also include competition involvement, a task which the company found most agreeable. The chassis number prefix AN5 identified the new Austin-Healey Sprite.

While MG busily produced the new Sprite from March 1958 at its Abingdon factory, Healey's vigorous and highly successful competition programme gave rise to a comprehensive list of options. In addition to normal factory options such as a heater, front bumper, laminated windscreen and rev counter, owners could now purchase all manner of Healey Motor Co parts to improve performance, braking and handling, while BMC also offered performance options and equipment from its own Special Tuning Department at Abingdon. The Sprite must have represented the first major target of the 'after-market' industry, as other companies soon introduced their own ranges of accessories and modifications.

During 1960, BMC instructed Healey to redesign the front end of the Sprite to counter criticism of its awkward bonnet, and to revitalise interest as sales were dropping. BMC then went on to ask MG to redesign the rear end to incorporate an opening boot lid, but omitted to inform either concern of what the other had been requested to do! Fortunately for all three parties, common sense won through so that both Healey and MG knew their respective ends would complement each other.

The Healey end was finished first and fitted to a development Sprite. Gone were the single-structure bonnet and 'Frogeyes', to be replaced by a more conventional bonnet with separate front wings and a grille panel bolted to the car's main framework. The headlights were now mounted in the front wings, while the previous smiling radiator grille gave way to a rather nondescript and expressionless rectangular grille. For MG's part, the tail was squared off and a conventional opening boot lid appeared. These modifications resulted in a Sprite which was far more practical to live with, but which had also lost something of the original's characterful appearance.

The engine remained at 948cc, but output was raised to 46.5bhp with the aid of twin 1¼in HS2 carburettors, larger valves with twin valve springs, a new camshaft and a higher compression ratio. For the first time, a choice of compression ratio could be ordered to suit local conditions. New bumpers, side and tail light units were fitted, and a close-ratio gearbox from the competition options list became standard. The new Sprite, now identified as the MkII (chassis prefix HAN6), was announced in May 1961 to leave room for BMC to sell off all remaining stocks of the MkI (as it then became known), which had finished production in November 1960.

To open further market potential, an MG version was announced in June 1961. Although very much a 'badge engineered' exercise, the MG Midget (chassis prefix GAN1) was a slightly up-market offering. Full waistline trim strips, a central bonnet moulding strip, vertical slat grille and a slightly improved interior added £15 to the Sprite MkII's price tag. Badging changed accordingly, but the Midget lost the passenger 'grab handle' fixed to the dashboard and wore plain chrome hub caps rather than the distinctive 'AH' embossed items of the Sprite.

October 1962 saw the arrival of disc front brakes and an increase in engine capacity to 1098cc (and 55bhp) for both Sprite (HAN7) and Midget (GAN2). The interiors were also improved with the addition of a padded roll to the lower edge of the dashboard, covered door top mouldings, better floor coverings and a new electric rev counter to replace the earlier cable-driven type. These improvements, however, did not prompt a change in designation.

Seeing the success of the Sprites and Midgets, Triumph in 1962 entered the small sports car business with the Spitfire, which was based largely on the running gear of the Herald saloon. No longer having the market to itself, BMC announced major revisions to the Sprites and Midgets in March 1964, the new models becoming the Sprite MkIII (HAN8) and Midget MkII (GAN3).

To counter the Spitfire with its winding windows, the perspex sliding sidescreens were replaced by completely new doors with winding windows, hinged quarterlights and exterior door handles; the doors could also be locked from both outside and inside. A more curved windscreen in a new frame was provided with a support strut in the centre which also carried the rear-view mirror, allowing the mirror now to be adjustable for height. A completely new dashboard with a small parcel tray on the passenger side was fitted, but the Sprite's added feature – the

Craig Polly's Sprite MkIV 'facelift' model in New Racing Green with Dave Gilbert's early Midget MkIII in Old English White.

grab handle – was discarded. The hood was revised with a full header rail at its leading edge to simplify removal and refitting to the windscreen frame by means of two overcentre clips operated from inside the car.

Although remaining at 1098cc, the engine was revised with a larger-bearing crankshaft and a new cylinder head with larger valves and much improved porting. Together with a new exhaust manifold coupled to a larger diameter exhaust system, these changes helped to raise power output to 59bhp at 5750rpm. The old mechanical fuel pump was replaced by an electric SU pump fitted above the right-hand rear shock absorber of all places! More surprising, however, was the fact that it was now deemed no more expensive to use full semi-elliptic springs instead of the former quarter-elliptic type. Improved ride quality and better lateral location of the axle were advantages, although axle tramp could now become a possibility under extreme conditions.

The Sprites and Midgets continued in this form until October 1966, when respective MkIV (HAN9) and MkIII (GAN4) versions were announced. Three essential changes took effect. First, the engine grew to 1275cc, a size shared with the famed Mini Cooper S although the 65bhp of the 'Spridget' engine was 11bhp short of the Cooper S's rather specialised unit. The need for cheaper and easier-to-manufacture components precluded using the same specification as the Cooper S, but at least a discerning owner knew that the potential was there to make the new Spridget a much quicker car. Second, the hood was much improved with a fully-folding frame which allowed quicker and simpler operation. To accommodate the new hood and its frame, the rear edge of the cockpit was moved back by 4in, giving a useful increase in storage space behind the seats when the hood was raised. Unlike its predecessors, the new hood was permanently attached to the bodywork along its rear edge, and a hood cover could be ordered to keep it neat and protected when lowered. Third, a completely new pedal box assembly with separate brake and clutch master cylinders replaced the single tandem unit of the earlier versions.

The largest single market for the Sprite and Midget was North America, where the arrival of new emission control and safety regulations on 1 January 1968 meant that simply transferring the controls from the right-hand side of the car to the left would no longer be sufficient. Initially, the new regulations could be met by adding an exhaust port air injection system to the engine, a padded facia panel, dual circuit brakes (with a dashboard warning light for a circuit failure), and a collapsible, energy-absorbing steering column. None of these modifications were applied to home market cars, which continued as before.

Emission and safety regulations became more stringent over the following years, causing a widening gap between the specification of North American and UK Sprites and Midgets. To add further confusion, European regulations also had to be met, but in general these were modest in comparison.

The 'Spridgets' continued to sell very well on both the home and export markets, but dark clouds were forming over the company making them. In January 1968, BMC merged with the Leyland organisation, makers of the rival Triumph Spitfire, to form British Leyland. The new management, of Leyland rather than BMC background, brought sweeping changes to the whole of the company's activities. The first of these to affect the 'Spridgets' occurred in October 1969, when facelifted HAN10 and GAN5 versions appeared with BL badges on both front wings.

The small but distinct differences between the Sprite and the Midget disappeared: the Midget's side trim strips, the vertical-slat MG grille and the Sprite's 'AH' hub caps were dropped in favour of a 'commonised' approach. Rostyle wheels of a fake alloy pattern were now the standard fitting, with wire wheels, an option since the arrival of the 1098cc engine, still being offered. A single design of radiator grille for both models was based on the simpler Sprite version but now finished in satin black, with a small central badge applied as appropriate. Bonnet badges were no longer fitted, while new style badges were fitted to the boot lid. Each black-painted sill bore Sprite or Midget lettering.

Initially the windscreen frame was finished in epoxy black but this proved to be unpopular. A new steering wheel with a covered rim and revised trim were changes to the interior. An extra silencer was fitted transversely behind the petrol tank to muffle the exhaust note. A slimmer front bumper was com-

This 'round-arch' Midget belonging to Denis Matthews is painted Mallard Green, a difficult colour to define as it seems to change under different light. The roll-over bar is a safety feature added by Denis. Number plates should be reflective white/orange from the 1972 L suffix.

The only one there is. Michael Cohen's 1975 Golden Jubilee Midget 1500 was built as a one-off by MG for presentation as a prize to a lucky winner from the company's workforce. The eventual winner turned out to be a non-motorist who then sold this Midget to a British Leyland dealership to use as a showroom display car. Total miles to date? Just 80! It remains to be seen whether the first free service will be honoured…

plemented at the rear by quarter bumpers, a square number plate fitting and new rear light lenses.

Altogether, this was a cosmetic job compared with the process of meaningful development we had witnessed in the BMC days. Prices also became commonised, the Midget no longer carrying a higher price for additions which, of course, it no longer boasted. At least a heater, previously optional, was now included in the price.

One may well have questioned by this stage why separate Sprite and Midget versions needed to continue in any case. The answer was just around the corner. At the end of 1970, the agreement between the now-defunct BMC and the Donald Healey Motor Co expired. Since Donald Stokes, the British Leyland supremo, did not share Leonard Lord's view that consultant companies had any part to play in corporate policies, Healey's agreement, like that of John Cooper for the Mini Cooper, was not renewed. Rather oddly, Abingdon did continue to build the Sprite, but hastily re-badged this as an Austin Sprite (AAN10) until July 1971, when production became purely MG Midgets. The export markets at least were spared any confusion, as neither the facelifted Sprite HAN10 or Austin 'only' Sprite AAN10 were offered outside the home market.

Towards the end of 1971, the stylists again made their mark by attempting to make the Midget look different with minimal investment, for the BL purse strings were very tight by this time, especially in the MG sector of its domain. For the 1972 model year, the Midget cheekily revealed all of its rear tyres with new rear wings which featured fully rounded wheel arches instead of the previous squared-off shape. A small flare to the edge of each wheelarch finished off the new look of what became known simply as the 'round-arch' Midget. A new pattern of Rostyle wheel, very similar to that used on the MGB or by Ford, replaced the 'fake' alloy wheel pattern, while wire wheels remained an option. Inside, rocker switches replaced toggle switches on home market versions, while petrol tank capacity was usefully increased from six to seven imperial gallons.

Owing to increasing legislation in the all-important North American market, where the Midget continued to sell well, more drastic changes were now required in order to keep pace. Added to the earlier modifications, safety bonnet hinges, head restraints, side marker lights, a third windscreen wiper blade, a seat belt warning system, a fuel evaporative loss control system, reduced compression ratio and impact-absorbing rubber buffer overriders all appeared on North American Midgets between 1968 and 1974.

Facing even more stringent emission control and 'low speed impact without damage' regulations, major revisions occurred in October 1974. The need for a catalytic converter and the ability to use unleaded fuel were on the horizon in North America, or, in the case of the Californian market which was generally a year or two ahead of Federal

13

standards, just around the corner.

The A-series engine simply could not meet the coming regulations without considerable development and expenditure. Instead, a cheaper solution was presented by the 1493cc engine of the Midget's greatest rival, the Triumph Spitfire. The Triumph engine proved relatively simple to adapt for the Midget's engine compartment, requiring only a new exhaust manifold and sump pan, and the transfer of wiring, piping and plumbing as all ancillary engine equipment was on the opposite side. The new gearbox, derived from that of the Morris Marina, at long last gave synchromesh on first gear, but conversely the gear ratios were not as good as before.

With a single Zenith-Stromberg carburettor, exhaust gas recirculation, exhaust port air injection, anti-run on valve, temperature-controlled carburettor air intake and 7.5:1 compression ratio, the North American specification engine managed just 55bhp. Even so, a catalytic converter and lead-free fuel were further requirements which had to be met immediately in California, so Triumph-engined Midgets initially could not be sold there until the necessary development work had been completed. This achieved, BL had to sell two different specification Midgets in North America, to meet Federal regulations and the more exacting Californian standards. If this were not enough, home market Midgets were using the 1493cc engine with twin SU 1½in carburettors and a 9:1 compression ratio, thereby giving *three* different engine specifications. Although Federal regulations moved on to demand lead-free fuel and a catalytic converter, Californian requirements also advanced, so there were dual North American specifications throughout the Midget's final phase.

The engine and gearbox were not the only items to be changed in October 1974. The low-speed impact and bumper-height regulations gave rise to large, black, polyurethane-covered bumpers at front and rear. The bodyshell also had to be strengthened, with the addition of new boxed frame rails in the engine compartment and boot to carry the new bumpers. The short-lived 'round-arch' rear wings were now displaced by the old square-topped variety as these were better able to resist damage from a rear end impact.

Owing to the extra weight and the need for the bumpers to sit at a prescribed height above the road surface, longer-travel springs with reset heights were fitted. Although these allowed good ride quality in relation to the increased weight of the engine and bumpers, the much praised handling of the previous versions was compromised. There were few, if any, favourable comments from drivers used to the nimble and predictable nature of the earlier versions.

Logically, the revised Midget should have become the MkIV, but this designation was never applied. While the owner's handbook still referred to this as a MkIII, no-one else appeared to recognised the revised Midget as such, most people simply calling it the Midget 1500 or 'rubber-bumpered' Midget. The chassis prefix became GAN6.

The Midget 1500 was to be the longest running version, and continued to receive minor specification changes for both the home and North American markets – by this time very few Midgets were exported to other countries. The final 500 home market examples off the Abingdon production line were all painted black, the very last one in December 1979 being retained by the company. It now lives in the British Motor Industry Heritage Trust collection at Gaydon in Warwickshire.

A total of 354,164 Sprites and Midgets were built between 1958 and 1979, with over 271,000 being exported. Exports represented 76.7% of total production, but not all examples arrived at their destinations in one piece.

Rather than suggesting damage in transit, I am referring to the Completely Knocked Down (CKD) kits of parts which were despatched from Abingdon to be assembled elsewhere. This system was a way round paying import duties in some countries on completed cars, and also proved a more cost-effective means of transporting cargo. Some receiving countries used locally-sourced components and materials to complete a CKD kit, as their legislation required a certain level of local content.

Small numbers of CKD kits went to Mexico, South Africa, Holland, Belgium and Eire (to Booth, Poole & Co Ltd), but by far the largest receiver of these was Australia. The Pressed Metal Corporation, a BMC subsidiary in Enfield near Sydney, started assembling Sprite MkIs in 1960 and continued without a break until 1967 with the HAN8 Sprite MkIII, but they were always behind with the model run compared with Abingdon's activities. Following a break, Sprite

The eyes of Innocenti are upon us. This is one of possibly only three Innocenti versions of the Sprite in Britain. Some 17,500 were built in Milan, using Sprite running gear in a Ghia-designed body. Neil Roger's example came to Britain via the USA in remarkably good condition. This model is the ultimate in luxury and styling in the 'Spridget' world.

SPRITE & MIDGET PAST AND PRESENT

production gave way to the assembly of Midgets, starting with the GAN4 in January 1968 and concluding with the 'square-arch' GAN5 in 1971.

Mechanical components were also supplied to Innocenti in Milan from October 1960 until 1970. Innocenti produced its own Ghia-designed bodyshell, but used standard Sprite/Midget running gear. Production updates affected the Innocenti, initially called the 950 Spyder, in parallel with the changes made at Abingdon. Thus when the 1098cc engine became available, the 950 Spyder was superseded by the Innocenti S. However, due to Italy's taxation system for motor vehicles, the 1275cc engine was never offered and BMC continued to produce the 1098cc engine specifically for Innocenti after October 1966. In 1967, the S was revamped by OSI, the bodyshell manufacturer, the most significant revision being to include a fixed-head coupé roof for the new Innocenti C Coupé. Around 17,500 Innocenti Sprite/Midget cars were built, chiefly for the home market but with exported versions reaching the USA and Switzerland. Although bodyshells were all built in Italy, normal Sprite chassis numbers were applied until Innocenti began to issue its own sequence of serial numbers in 1964.

The final non-Abingdon assembly plant was no more exotic than 'up the road' in Cowley. Between January and March 1967, a few early batches of Sprite MkIVs (HAN9) and Midget MkIIIs (GAN4) were assembled here. The Cowley plant was already involved in the Sprite and Midget story as paint was applied to the bodyshells here on their journey from manufacture in Swindon to assembly in Abingdon. Although no official reason was given as to why 489 Sprites and 476 Midgets were assembled at Cowley, a possible theory is the fact that an engine problem halted deliveries shortly after announcement of the new 1275cc versions. Cowley was quite likely employed to catch up with production once improved engines became available in early 1967. There is another view that Cowley was being considered as a permanent home for 'Spridget' production, as Abingdon's future was beginning to seem precarious.

The entire range of Sprites and Midgets, no matter which version or where assembled, filled a niche in the market that has never subsequently been matched in such a successful manner. The cars brought the joys of economical sports car ownership not only to more than 350,000 customers, but also to possibly something like 10 times this number of subsequent owners as they changed hands so frequently. Short-term ownership was quite common: once bitten by the sports car bug, owners often moved on to bigger and faster things. But better? Judging by the number of people who have told me "I used to own one of these" and added that they regret selling it, Sprites and Midgets are remembered with a depth of affection and fondness seldom afforded to other vehicles.

Bob Barker's late Midget 1500 – one of a batch of 500 cars painted black – has covered a much higher mileage than the Golden Jubilee Midget, for it is approaching 1000 miles from new. There is no British Leyland motif on the front wing: this was dropped at an unknown point during production.

SPRITE MkI (AN5)

Following on from the success of the partnership agreement between Austin and the Donald Healey Motor Co Ltd to produce the Austin-Healey 100 series of sports cars, a completely new car, the Sprite, was announced on 20 May 1958.

As before, Austin would supply the main components and manufacturing facilities while Healey handled the design and development work. As part of the giant BMC organisation, Austin used a number of its manufacturing sites to produce the components for the new Sprite. Bodyshells were pressed and welded at Swindon upon floorpan units supplied by John Thompson Motor Pressings of Wolverhampton. Paint was applied to the bodyshells at Cowley before they were passed on to the MG factory at Abingdon for final assembly. The slightly modified Austin A35 engines were assembled at the Morris Engines plant at Coventry. The many other components required to make the Sprite were delivered to Abingdon from other BMC plants and numerous sub-contract companies.

Although the Austin A35 was the main provider of basic parts for the Sprite, the Morris Minor parts bin yielded a rack and pinion steering system and individual hydraulic rear brakes. For the Sprite to be a true sports car, of course, twin carburettors were absolutely essential, so the A35's single Zenith was replaced by two 1⅛in H1 SU carburettors.

Part of the Sprite's appeal was a pure and simple approach which gave the driver just what a sports car should offer – fun. The Sprite also contained a special charm that was to earn it many friends throughout the world. The simple styling was broken by the positioning of the headlamps in fixed pods set into the top of the bonnet. Hence the car became more commonly referred to as the 'Frogeye' – or 'Bugeye' in America – Sprite.

Simplicity at the rear end left the Sprite owner with no opening boot lid. Instead, both seat backs could be hinged forwards, allowing the owner to gain some useful potholing practice in the dark chasm behind. Despite the poor accessibility, boot capacity was in fact quite good, as was the space afforded to driver and passenger. Without winding windows, the hollowed-out door frames offered good elbow room and useful storage pockets in the lower portions. Detachable sidescreens provided access to the interior door handles, for no exterior handles or locks were fitted. Security? Only an ignition key was provided: nothing else could be locked unless the optional lockable petrol cap was ordered.

The Sprite was to prove a highly successful competition car in both racing and rallying, the Healey Motor Co Ltd developing a huge range of modifications and equipment to tempt owners who sought the very best from their Sprites. Many other companies, too, were quick to see the Sprite's potential by offering their own performance and personalising goodies. Indeed, for any Sprite to remain in absolutely original specification – as it left the factory gates – for any length of time became something of a rarity.

Not all Sprites, however, left the Abingdon factory in one piece. BMC's practice of supplying Completely Knocked Down (CKD) kits of parts for final assembly by overseas subsidiary companies, in some cases to avoid import duties, was also applied to the Sprite. Most CKD Sprites went to Australia, but, rather interestingly, CKD kits sent to Italy were built into completely different bodyshell units by Innocenti of Milan.

The cheap and cheerful Sprite – recognised by the chassis prefix AN5 – contained a cheeky, characterful ingredient which made it an instant success. Whether by design or by accident, its good points have combined with its shortcomings to produce an overall appeal which has stood the test of time.

BODY & CHASSIS

The Sprite was to be the first mass-produced British sports car to feature a combined body and chassis unit, known as unitary or monocoque construction. This simple structure had to be capable of withstanding all the stresses placed upon it without a separate, sturdy chassis frame for support. Although this type of construction was becoming well developed in saloon car design, a 'roofless' structure had to be sufficiently strong in the floorpan to prevent scuttle shake and to provide beam and torsional rigidity.

To achieve the necessary strength, much use was made of enclosed box sections, these being carefully arranged to accommodate all the loadings imposed by the suspension, steering and power unit. The entire structure, made up of some 200 sheet steel pressings, was welded together using spot welds and seam welds in the areas subjected to the heaviest loadings.

The main floorpan was virtually one flat pressing, the only shaping being the footwell depressions and

Owned by Allan Cameron, this beautifully preserved Sprite MkI has covered just 7600 miles from new. The lack of exterior door handles has confused many an unfamiliar 'Frogeye' passenger. Handles, of course, were provided inside. Wing mirrors were a BMC optional extra.

Chunky but cheerful, the rear of the Sprite MkI (left) was kept beautifully simple. The standard non-lockable fuel cap was painted body colour. The hardtop is the BMC optional type and followed the style of the Healey version – Ivory White was the only colour available for the hardtop.

SPRITE MKI

Original Sprite & Midget

The face of fun. How could anyone not love it?

five transverse flutes either side of the centre section to provide strength beneath the seats. Spot-welded to the centre of the floorpan was the deep-section transmission tunnel which formed an enclosed box for much of its length. To each outboard edge of the floorpan, a full-length, deep-section sill provided beam strength. This was made up of a vertical inner panel with an outer panel spot-welded to it to form a 5in deep rectangular box section. The front portion of each sill swept upwards under the front wing to form a 7in deep section at the leading edge behind the front wheel. No internal (or diaphragm) panel was enclosed in the sill. A large transverse box connected the sills and transmission tunnel at the rear of the floorpan, this section also containing stiffeners and the location boxes for the rear suspension quarter elliptic road springs. The springs were secured at their forward ends into the outer ends of the transverse box section or rear bulkhead. A thick-gauge steel plate beneath each spring location helped strengthen the floorpan and so acted as the main anchor plate for the spring attachment. Although quarter elliptic springing simplifies some aspects of design, there were additional factors to be considered.

The loadings to the single attachment points were far higher than in a semi-elliptic springing system, whereby part of the loading would be fed into the spring's rear attachment point to the floor. For the Sprite, the rear bulkhead had to be further strengthened immediately in front of the location box. This was achieved with two pressings of 'top hat' section, one near to vertical welded to the front face of the rear bulkhead (heelboard) panel, the other, running from its base, longitudinally along the floorpan beneath the seat. The front end of this formed a right-angled junction to yet another box section which ran transversely from the sills, through the transmission tunnel directly in front of the seats. This section provided a jacking point at each of its outer ends, each jacking point being a welded-in steel tube poking through the inner sill panel and protected by a removable rubber bung in the outer sill panel. The centre section of the transverse box provided the mounting point for the rear of the gearbox.

Also forming a junction with the transverse box were two more longitudinal boxed rails running forwards either side of the gearbox and engine to the very front of the car. These would then carry the front bumper mounting brackets (an optional extra on home market cars) and the bonnet closing locations. Two vertical pillars from these supported the radiator directly in front of the engine. The entire front suspension and steering gear, as well as the two engine mounting points, were accommodated on a transverse crossmember welded to the same boxed rails. Three welded nuts allowed each engine mounting bracket to be secured, while three more welded nuts on the top platform of the crossmember allowed each front lever arm shock absorber to be secured by ⅜ UNF bolts which passed through the shock absorber body.

Rising from the sills and transmission tunnel was the front bulkhead assembly containing door pillars, footwells, bonnet hinges, facia location and master cylinder locations for both left- and right-hand drive. A platform immediately above the gearbox bellhous-

Prior to the introduction of a full box-section junction between inner rear wheelarch and axle tunnel, early versions of the Sprite MkI had these gusset pieces to strengthen this critical area. The very first Sprites did not have the front gusset (on the left), these cars subsequently being recalled to have this fitted.

18

Proprietary Lucas lamp units, as found on a number of other contemporary vehicles. The Sprite badge is correctly finished in gold with red lettering. A chromed badge would be a modern reproduction. Fuel cap is the BMC optional lockable type made by Wilmot Breeden.

The rubber buffers, screwed to the sills, are often missing when new sills have been fitted. These buffers provide a seating for the front wings when the bonnet is closed. Restorers should not forget to measure the position of the screw holes as new sills are not pre-drilled.

ing provided space for the single 12-volt battery and Smiths heater box. Hot air from the heater flowed downwards into a plenum chamber and rearwards, via small 'letterbox' flaps, into the inner footwell panels, while separate ducting took warm air to the demister slots set into the top of the bulkhead. To the front of the footwells, the inner front wings rose up over the front wheels, a steel bracket connecting the front edge of each inner wing to the radiator support pillars to prevent the wings from vibrating.

Returning to the rear of the car, the rear bulkhead, which was some 10in (255mm) high at its forward edge, also formed the leading edge of the boot floor. This ran over the rear axle tunnel and then fell away to form the boot floor. The fuel tank was mounted directly beneath this floor, while the spare wheel lay flat on the floor and was secured by straps. A triangular pressing ran down each side from the inner rear wheelarch to taper out at the rear of the boot space, these pressings being the main internal supports for the boot floor. From the rear of the bulkhead and parallel to each rear spring on the inboard side, a large triangular gusset plate further supported the leading edge of the rear axle tunnel and provided a suitable mounting for the lever arm shock absorbers and radius arm brackets.

The rear bodywork of the Sprite MkI comprised only three outer panels: two rear wings and a centre 'shroud' panel which extended down from the rear of the cockpit to join the boot floor. No boot lid opening was provided, access to the boot instead being made through the cockpit. The shroud and wings were spot-welded together along the top edges of each wing, and a beaded filler strip was sandwiched into the joint. This was an expediency to save filling and blending this long join line in production. The rear light and indicator mountings were incorporated on this join line as it fell away to its lower edge.

Fairly early in production it was realised that the monocoque lacked strength in two particular areas. Indeed, the first cars produced were subsequently subjected to having an additional strengthening bracket welded into the rear inner wing where this met the axle tunnel. The brackets were soon included in production, but from AN5-4333 (September 1958) a full box-section strengthener was introduced which ran the full length of the junction between inner wing and axle tunnel to extend rearwards to the boot floor.

To strengthen the junction between the inner front wings, front suspension crossmember and bulkhead, the following revisions were introduced from AN5-5477 (October 1958). An additional vertical panel was included in the engine compartment, effectively double-skinning the inner wing panel between the front suspension crossmember and bulkhead. Both inner and outer panels, spaced approximately 1in (25mm) apart, were produced with a large triangular aperture, a drain hole being provided at the lowest point as the cavity created would fill with water and debris thrown up by the front wheel. Where this new boxed section rose to meet the bulkhead, a gusset in the form of a quarter-circle was added to further strengthen the junction. The gusset on the right-hand side also served as the revised mounting point for the ignition coil, a small top-hat section being welded to the gusset for this purpose. Accompanying these modifications, the splash shields in the bottom of the engine compartment were revised, the right-hand shield no longer being removable.

The other areas to receive strengthening were the joins from the inner rear wing to the boot floor and rear bulkhead. A new box section began at the forward edge of the inner rear wing to continue along the entire join line of the wing to axle tunnel and boot floor. This terminated at the triangular panel running down from the top of the wing to the boot floor. The first few hundred cars off the production line were subject to a Technical Bulletin instructing Austin dealers to have a strengthening bracket welded into the inner rear wheelarch to reinforce this weak area. This was subsequently applied on the production line, with the full box section cornerpiece taking effect from AN5-4333 (September 1958).

The doors were of very simple construction, with the main door frame covered by an outer door skin which was crimped over the frame's flanged edges. The leading edge of the frame carried two hinges which were screwed to both the door and the door pillar, these screw mountings allowing adjustment to the door fit. The rear edge of the door carried the door lock mechanism which engaged on an adjustable conical catch screwed to the B post. The top edge of the door was finished with a bright anodized aluminium extrusion to match that fitted to the rear cockpit edge, the door finisher being drilled to take two screws to retain the sidescreen frames.

The rear-hinged bonnet was a large construction comprising a centre section, headlamp pods, front valance, radiator grille and both front wings. Each of the two pressed steel hinges was secured to the bonnet by four ¼ UNF screws into welded nuts in the rear crossmember. The rear of each hinge fitted into a sunken box in the bulkhead panel and pivoted about a ⁵⁄₁₆in diameter bolt which passed through the box and hinge. As at the rear of the car, a beaded filler strip was used to finish the join between wings and centre section. To further brace the front wings to the centre section, a simple 9in (228mm) long bar was bolted between the rear crossmember and the front wing trailing edge stiffener. From chassis number AN5-4695 (September 1958), ducting panels were added behind the radiator grille to improve air flow into the radiator.

Seat belt mountings were not included, but all bodyshells were completely equipped for either right- or left-hand drive applications. Screwed-on blanking plates or rubber bungs covered whichever holes were not required.

BODY TRIM

Exterior trim was kept to an absolute minimum, with a distinct lack of unnecessary adornment.

The chrome-plated radiator grille, a one-piece pressing, was secured to the bonnet by spire clips. Beneath this, a chromed T handle locked the bonnet shut but did not contain a key-operated locking barrel. The handle operated two transverse locking rods which engaged in locking plates bolted to the front chassis rail extension. The T of the handle would point across the car when in the closed position and turned through 90° to release the locking rods. The rods and their operating lever plate were painted body colour, the plate being held to the bonnet by four crosshead 10 UNF screws. A safety catch was provided to stop the bonnet from lifting if the locking rods were not engaged, although in practice the bonnet was so heavy that it was unlikely to blow open at speed. The catch, a simple spring-loaded, black-painted handle on the left-hand chassis rail extension, engaged in a cut-out section in the bonnet lower ducting panel.

The bonnet was held in the open position by two nickel-plated struts which engaged automatically. To release these, the bonnet had to be lifted slightly before lowering. As a further bonnet support, a fixed-length rod pivoted from a bracket on the right-hand inner wing support bracket and was stowed in a spring clip on the left-hand bracket when not in use. The end of the rod was bent to engage in a hole in the bonnet's right-hand duct panel. The rod and the

The front bumper was an optional extra for home market Sprites but a standard fitting for export. The number plate was attached to a black-painted backing plate.

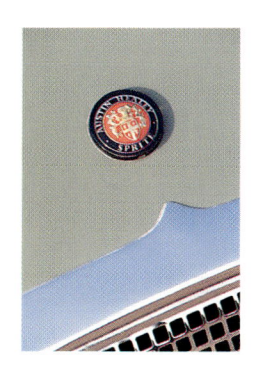

Perhaps the most elaborate feature of the Sprite MkI was the bonnet badge containing the heraldic Austin motif. The radiator grille was a single-piece pressing.

On a car without a bumper, the number plate was mounted on a fabricated steel plinth. Four nuts and screws secured the plinth to the front valance, which was always pre-drilled to accept it. Paint finish for the plinth seems to be a matter of conjecture: black or, possibly, body colour. The bonnet release handle, somewhat obscured by the plinth, contained no locking barrel. This handle is a modern reproduction.

support brackets were painted black. When closed, the bonnet side wings rested upon rubber buffer blocks, two on each side, screwed to the curved top face of the sills. From AN5-6063 (October 1958) these were augmented by two further rubber support blocks mounted on the top of the inner front wings and sitting on aluminium spacers. These extra blocks also became a retrospective fitting by BMC dealers when owners complained that the bonnet was resting on the engine!

The front bumper was listed as an optional extra on home market cars but was standard on export versions. The chromed bumper blade carried two vertical overriders with a black plastic seating moulding sandwiched between the mating edges. The front number plate was also carried by the bumper on a black-painted steel backing plate, the size of which varied according to local regulations. The bumper was mounted by four black-painted steel brackets to the front frame rails and secured by two ⅜ UNF bolts into welded-in bobbins in each rail. If no front bumper was fitted, the number plate would be carried on a black-painted (or possibly body-coloured) fabricated steel plinth attached to the lower valance by four screws.

No rear bumper was fitted as such, all MkI Sprites having two vertical overriders which, rather curiously, did not match the pattern of the front overriders (when fitted). Each rear overrider was attached to the body by two black-painted steel brackets: the upper bracket formed an inverted U with a rubber moulding sandwiched between the body and bracket, while the lower bracket was bolted by two 5⁄16 UNF bolts through the boot floor into a strengthening piece inside the boot. This strengthening piece formed a triangular corner between the boot floor and rear panel, and contained welded-in nuts for both top and bottom overrider brackets. The black-painted strengthening piece was in turn bolted to the boot floor from underneath by one 5⁄16 UNF bolt.

The rear number plate was screwed directly to the body without a mounting plate, a recess being formed in the rear panel to provide a flat mounting surface. The number plate lamp was mounted centrally above the number plate on a body-coloured steel plinth, this being mounted on a black rubber pad and screwed to the body from inside the boot area.

The windscreen was generally of toughened glass, but laminated glass was a listed option for the home market and standard for most export markets including North America. The screen was mounted into a chrome-plated brass frame with a rubber seating. The lower edge of the frame contained an inverted T slot to take the frame-to-body sealing rubber strip.

Either side of the frame, a pillar – body coloured on the earliest cars – provided support and was secured to the top of the door pillar by two ⅜ UNF countersunk screws. The front screw was retained by a captive nut within the pillar, the rear screw having an exposed nut which was accessible with the door open. A rubber pad was sandwiched between the pillar and body. The screen frame was slotted into the pillars with three 10 UNF screws locking the frame into each pillar. The pillars and frame were revised at chassis number AN5-5477 (October 1958), the new alloy pillars, now unpainted, having a more rounded section. The top of the new anodized alloy frame was modified to take an improved method of fitting the hood, with a slotted leading edge. The centre seven 'lift the dot' fasteners were deleted, only the outer fastener on each side remaining.

The sidecurtains were fastened to the top edge of each door initially by two large-headed ¼ UNF screws, but from AN5-1606 (June 1958) the captive nuts in the door and the holes in the aluminium moulding were enlarged to accommodate 5⁄16 UNF screws. The head of each screw contained a large slot, sufficient to enable a coin to be used to tighten or loosen the screw, and the edge of the head was knurled to assist finger grip once the screw had been loosened. All four screws were chromed-plated and engaged with captive nuts in the top of door frame. Slotted brackets at the bottom of the window frame enabled a sidecurtain to be removed or refitted without having to remove the fastening screw completely.

The two windscreen wiper spindles protruded from the scuttle panel immediately in front of the lower frame-to-body seal. A chrome plated Mazak ferrule controlled the angle of the spindle. The position of the spindles remained the same for either right- or left-hand drive, but the blade parking position was to the right for the right-hand drive and to the left for left-hand drive. The 8in (203mm) Trico Rainbow blades were attached to chrome-plated blade carriers with hook connection to the chrome-plated arms. The jets of the optional windscreen washer would be positioned between and forward of the spindles, although the exact position could vary slightly if dealer fitted; some instances of bonnet-mounted jets are not unknown. The Tudor windscreen washer used two separate jets, while the Trafalgar washer kit used a single jet head with two outlets.

Badging was confined to a bonnet badge and a 'Sprite' motif on the rear shroud panel. The bonnet badge, of 2⅛in (58mm) diameter, sat centrally in a hole above the radiator grille, and was secured by three clips. The badge carried the heraldic Austin logo picked out in red and chrome. Surrounding this, chromed lettering – 'Austin Healey' at the top and 'Sprite' at the bottom – appeared on a black background. The badge was made of clear plastic with all colouring being applied from the underside. The rear badge, positioned on the left-hand side above the number plate recess, carried Sprite script set upon an arrow. The arrow and script surround were painted gold with red for the inset lettering. Made of Mazak, the badge was secured by two spire clips from within the boot space.

WEATHER EQUIPMENT

The hood, made from ICI Vynide material with AS6 Vybak windows to the rear and rear quarters, was initially available only in black. From chassis number AN5-5477 (October 1958), the method of securing the hood to the windscreen frame was revised to improve weather sealing and reduce wind noise. The nine 'lift the dot' fasteners were replaced by a steel strip located into a small pocket formed along the leading edge of the hood material, this steel strip engaging with a slot in the underside of a revised windscreen frame top rail to provide a continuous seal along the width of the head. A single 'lift the dot' fastener remained each side to provide sideways location and to add tension to the hood. The option of a white hood became available from AN5-13543 (April 1959).

The rear of the hood had a pocket with an inserted steel strip similar to that at the front, but cut away in two places to expose short lengths of the steel. These exposed portions engaged with corresponding slotted anchor plate escutcheons mounted on the rear shroud. The chrome-plated escutcheons also served to locate the optional tonneau cover in the same manner. Two Tenax fasteners outboard of each escutcheon and a turn-button fastener just behind each door completed the attachment of the hood to the body.

The hood was held erect by a frame (always painted Cumulus Grey) consisting of two tubular bows hinged together with flat links and pivots to allow the frame to fold flat for stowage. The rear bow contained a spring-loaded lower section which engaged with a location tube immediately behind the door B posts. The spring-loaded section was provided to tension the hood in the raised position, but to facilitate raising and lowering the spring could be locked into the fully compressed position by rotating the lower section of the bow and engaging a peg into a slot. A lug was provided on the bow to allow a means of rotation.

Once the hood was erected, the lug plate could be rotated to release the peg and to allow the springs in each rear bow to tension the hood. Hood frames for the nine-stud hood (before AN5-5477) did not contain the spring-tensioning facility. When not in use, the hood frame would be hinged into the flat position and removed from the B post location tubes. The frame could then be turned through 180° and fitted into small location boxes screwed to the heel panel behind the seats. The top of the frame bows would fall just inside the lower edge of the rear shroud, and a strap hanging from a slot in the shroud stiffener rail would then secure the bows and prevent

Ian Wilson's early Sprite MkI has the appropriate nine-stud windscreen frame and hood. Primrose Yellow paintwork was discontinued in February 1959.

Fabric-covered side curtains with Vybak windows were standard equipment without a hardtop. Button fasteners could be released to gain access to the car or for hand signals. The complete frame could be removed from the door by releasing two screws. The dashboard clock is not original.

Aluminium-framed sidescreens with a sliding perspex back window were standard with a hardtop. These also became an optional alternative to the fabric side curtain until between March/May 1960, when all Sprites adopted this type of sidescreen as standard equipment.

SPRITE MK I

The hood was available only in black until AN5-13543 (April 1959), when a white hood became an option for some body colours. Note the stuck-on perspex blocks which served as 'handles' on the rear sidescreen pane.

The optional tonneau cover, initially available only in black, was supplemented by a choice of a white cover for some colour schemes from AN5-13543 (April 1959). The rear of the cover was retained by the same fasteners as the hood, while three 'lift the dot' fasteners at the front engaged with corresponding studs on the scuttle. The central zip allowed the driver's side to be opened while leaving the passenger side safely covered.

them from rattling around behind the seats. The owner's handbook demonstrated the method of folding and rolling the hood, before securing it with further straps to the bows in the stowage position. In practice, to fold and stow the hood in this manner was both time-consuming and damaging to the hood and its Vybak windows. Most owners more sensibly folded the hood between the window joins and stowed it flat inside the boot, space and luggage permitting.

Initially, simple side curtains consisted of a Vybak sheet edged with black Vynide formed around a steel frame which engaged with two chromed screws on the top of each door. The lower edge of each curtain could be released from the frame to gain access to the internal door handles or to allow hand signals. From AN5-1606 (June 1958), the side curtain retaining screws were increased from ¼ to 5/16 UNF, necessitating revisions to the frame brackets, door top mouldings and captive nuts. The side curtains could be completely removed by slackening the two screws and sliding the side curtain frames forwards and clear of the retaining screws. From chassis number AN5-13543 (April 1959), white side curtains became an option to complement the newly introduced white tonneau cover and white hood.

From chassis number AN5-8283 (December 1958), an optional fibreglass hardtop became available in Ivory White only. With this came new sliding pane sidescreens in aluminium frames. These fitted in exactly the same way as the curtains, but provided better weather-proofing and visibility. Only the rear of the two perspex panes was free to be moved, the front pane being crimped into the frame. A small rectangular block of perspex stuck to either side of the rear pane acted as a very small handle. Surrounding the aluminium frame, a rubber moulding provided further weather-sealing against the door, screen pillar and hood.

These sliding pane sidescreens became an optional fitting without the hardtop being ordered, but from AN5-34558 (March 1960) they became listed as standard, although a few cars still had the original sidecurtains until AN5-39945 (May 1960). Sidescreens of different shapes and dimensions became available to suit alternative hardtops from other manufacturers. The standard sidescreens were supplied by Weathershields, as were hoods and tonneau covers.

The optional full-length tonneau cover with centre zip was available either in black or, from AN5-13543 (April 1959), white. This used the same combination of fasteners at the rear, but three 'lift the dot' fasten-

The Weathershields hood frame, correctly finished in Cumulus Grey, did not have the spring tensioning portion in the upright for the early 'nine-stud' windscreen versions. Note the yellow piping on the black seat, only found with Primrose Yellow paintwork.

23

ORIGINAL SPRITE & MIDGET

The 'official' way to stow the hood! Few owners found the time to fold, roll and strap the hood in this manner. It was much quicker just to throw it in the boot – and quicker to make ready to re-erect when the heavens opened up too! The narrow strap shown here is a modern replacement. Note also how the door draught excluder is tucked down behind the trim liner panel. This may look an odd means of securing the end, but it is the way BMC intended. A screw passes through both panel and excluder.

ers spaced across the scuttle panel behind the windscreen provided the front fixings. The middle fastener was positioned to the left of the rear-view mirror on right-hand drive cars and to the right on left-hand drive cars, enabling the driver's side of the tonneau to be unzipped and stowed behind the seat while leaving the passenger side covered.

Although stowage bags were provided for the sidescreens, the tonneau cover was simply folded and placed in the boot space. The boot housed the spare wheel, which was left uncovered but secured flat against the floor by webbing straps. The three legs of the strap engaged with three hooks welded to the boot floor, two at the back and one centrally at the front.

Also to be found in the boot was the tool kit, which comprised a Smiths 'Steady Lift' side jack and pressed steel ratchet handle, hub cap remover, wheel brace, spark plug box spanner and tommy bar, all stowed in a black vinyl roll. From AN5-42200 (June 1960), this roll was changed and, it would appear, the option of additional tools was removed from BMC's accessories list (see 'Options & Accessories' later in this chapter).

INTERIOR TRIM

The two bucket seats each comprised three main components: a Dunlopillo cushion on a pressed steel base, a hinged backrest and an L section base frame. The cushion rested in this black-painted frame, while the seat back was fitted via a welded 5/16 UNF bolt in a steel strip rising from either side of the frame to provide the hinging required for access to the boot.

Only the driver's seat could be adjusted for reach, by means of sliding seat runners under the base frame. The floor-mounted runners were pierced with square holes, a spring-loaded locking lever engaging with a hole to secure the seat into the desired position. The lever was pivoted to the upper runner and operated from the outboard side of the seat. The floor runners were mounted on 3/8in (9.5mm) thick packing strips of wood, which gave the seat frame sufficient clearance above the longitudinal 'top hat' box member running from the rear heel panel to the jacking point box section. From AN5-26139 (October 1959) the runners were revised to allow further forward adjustment of the seat. Since the passenger seat differed in having no adjustment beneath the frame, four brackets were either bolted or, from chassis number AN5-31902 (January 1960), riveted to the frame to give a matching seat height. Both seats were secured by four 1/4 UNF screws into welded nuts.

The seat coverings were Vynide and coloured according to the exterior colour scheme, as listed at the end of this chapter. The cushions and backrest faces were fluted with contrasting coloured piping around the outer seams. The seat material was secured to the frame by steel clips driven over the material and seat pressing edge. All interior trim panelling also consisted of Vynide, colour-matched to the seats and laid over hardboard panels which were held in place with chromed self-tapping screws in cupped washers.

The panel applied to the door differed in that the material was laid over an additional thin cardboard former to produce a shallow recessed effect in the centre of the panel. The top edge of this was further formed by a curved steel edging strip which produced a convex panel allowing an increase in the

The revised Weathershields hood frame with spring-tensioned bows, to accompany Sprites with the 'two-stud' windscreen.

Seat coverings were Vynide on a Dunlopillo base and padded backrest. Old English White Sprites were the only ones to have a choice of interior trim colour, black being offered in addition to red from January 1959.

24

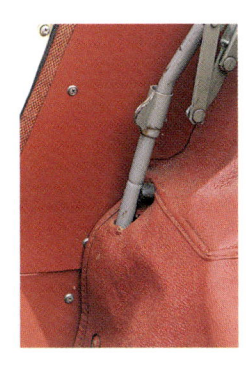

SPRITE MK I

The hood frame – this is the later spring-tensioned type – in the stowed position in the lower location boxes beneath the matting. Note the thumb plate and rotating lower portion that contained the coil spring.

door pocket space. Steel strips applied to the other three edges of the panel slotted over the door frame aperture to locate the panel. The panel was secured to the frame by two screws and nuts at the top, and two pop rivets lower down. Four rubber plugs, marked C-45 in raised characters, covered the hinge screw access holes in the front of the door.

The door tops and rear cockpit edge were finished with bright anodized, extruded, alloy mouldings. The scuttle edge, above the dashboard, was also finished with an alloy moulding but covered in trim colour Vynide. All finishers were secured with 10 UNF countersunk crosshead screws and nuts. The inside face of the door frame was painted body colour. The inside of the door skin was covered in ribbed rubber stuck directly to the skin, this again being colour-matched to the trim. A fabric check strap in trim colour prevented the door from opening too far and striking the A post skin.

The door catch was operated by a lever protruding from the top rear edge of the door frame. From chassis number AN5-10344 (January 1959), the lever was shortened and a chrome knob added. A rubber draught-excluding moulding, which included a fabric inside facing to match the interior trim colour, was clipped to the flange edging around the door aperture. A metal P clip secured the top of the draught excluder to the outer screws of the scuttle trim moulding rail. At the rear, the draught excluder was simply tucked down behind the trim liner panel

The early type of long door handle and fabric side curtain.

Deep door pockets provided some useful stowage space with good elbow room above. A steel edging strip gave the Vynide-covered hardboard panel its convex shaping. Note the short door handle with a chromed knob, fitted from AN5-10344 (January 1959).

and secured with a self-tapping screw.

The original floor covering, extending to the transmission tunnel, was thin-ribbed rubber matting colour-matched to the interior trim, but this is no longer available. Service life of this matting was fairly short, with bleaching of the colour and cracking being common. The matting was inscribed 'Lift to Fill' over the rubber blanking plugs on the left-hand side of the transmission tunnel to remind owners that access to the gearbox oil filler plug could not be achieved any other way. The portion of matting over the longitudinal top-hat stiffener beneath the seats was deleted on later cars. This change took effect according to trim colour at AN5-21546 (red), 26185 (blue), 26464 (black) and 37492 (green), between August 1959 and April 1960. Restorers today are obliged to fit carpeting to cover the floor and tunnel. The domed steel cover around the base of the gear lever was painted body colour and secured by four self-tapping screws hidden beneath the surrounding rubber matting.

The inner rear wheelarches and the rear axle were covered with padded Vynide material called Hardura, of interior trim colour and finishing at the highest point of the tunnel. Within the boot, thick grey cardboard trim panels covered each side of the boot space, extending to the very rear of the boot. The boot floor was covered by a further piece of Hardura, again in interior trim colour. Prior to AN5-4333 (September 1958), the boot floor covering was box shaped and no side trim panels were fitted.

The crowning glory of any restoration to genuinely original specification would be these moulded floor coverings in ribbed rubber. Not renowned for lasting and unobtainable for many years, these are a rare sight indeed. The mat over the gearbox tunnel has the words 'lift mat to fill' moulded into it.

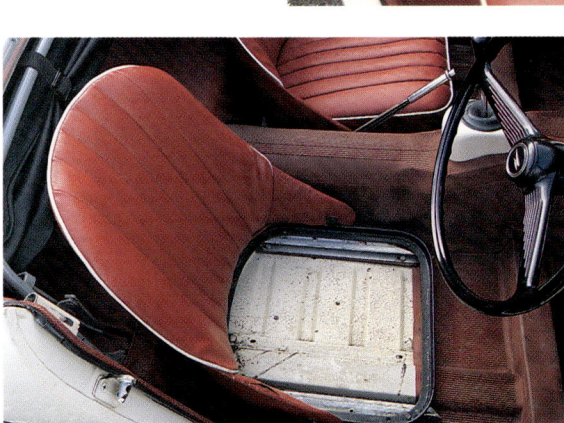

The sliding seat runners to provide adjustment for the driver's seat rested on wooden packing pieces, in order to raise the frame clear of the longitudinal top-hat section stiffener which feeds rear suspension loadings along the floor to the jacking point crossmember. The non-adjustable passenger seat used fixed brackets which did not require the wood strips.

Within the dark chasm, illuminated by flashlight. The luggage space was fairly generous, but any items disappearing behind the spare wheel demanded a long stretch in the dark to retrieve. Note the grey cardboard side liner panels and primer finish to the inside of the shroud panel.

Entrance to the dark chasm. With seats hinged forwards, access to the boot, which lacked a conventional lid, was restricted by the axle tunnel. The hood frame was secured by a central strap which fitted through a slot in the stiffener section under the shroud edge. Two further straps around the bows then retained the rolled-up hood.

DASHBOARD & INSTRUMENTS

The facia panel was made up of a single steel pressing covered in Vynide, colour-matched to the interior trim. The panel was secured to the scuttle by two ¼ UNF nuts and bolts and a central ¼ UNF stud welded to the scuttle stiffener. Two black-painted lower support strips connected the lower edge of the facia panel to the bulkhead.

On right-hand drive Sprites, the layout of the white-on-black Smiths instruments was as follows, from left to right (Smiths part numbers in brackets): combined oil pressure (upper) and water temperature (lower) gauges (GD 1502/01); rev counter (optional but very rarely omitted) with red main beam warning light (RN 2351/02); speedometer with total/trip mileage recorders and red ignition warning light (SN 6155/10); fuel gauge (FG 2530/31). A green flashing indicator warning light was fitted between the rev counter and speedometer, immediately above the steering column. The trip reset was placed directly beneath the speedometer in the lower edge of the panel. To the right of this, a sliding switch controlled the instrument illumination. For left-hand drive cars, the left to right displacement was as follows: fuel gauge (FG 2530/31); speedometer (SN 6155/10 or 6155/11); rev counter (RN 2351/02); oil/water gauge (GD 1502/01).

The speedometer and rev counter each had a silver centre section, the rev counter having a shaded portion between 5500rpm and 6000rpm. The needle spindle hubs were chromed, the needles being black with white end sections. The speedometer, reading to 100mph or 160kph, contained two numbers: 1472 for the number of cable revolutions per mile (920 per kilometre) and the Smiths 'SN' number which identified the instrument application as SN 6155/10 (mph) or SN 6155/11 (kph), both for a 4.22:1 axle ratio. The cable-driven rev counter carried the number RN 2351/02 and the marking multiplication of ×100 followed by the cable gearbox ratio of 3-1. Cars without a rev counter had a blanking plate over the aperture on the panel, and two red main beam warning lights set in a chrome bezel in the blanking plate.

The combined oil pressure and capillary tube sensing temperature gauge again carried white needles but with a black portion near the axis of each needle. Calibration was in lb per square inch to 100 and degrees Fahrenheit to 230, there being no metric alternative. The fuel gauge was more internationally understandable with a simple 0-¼-½-¾-F marking and a white needle pivoting from the top.

All instruments were surrounded by a chrome bezel and illuminated only when the main lighting switch was operated. There was, however, an independent sliding switch in the lower edge of the panel to extinguish the instrument lighting when the side and headlights were on.

In the centre of the dashboard, the key-operated ignition switch was surrounded by a rotating black knob for side and headlight operation. Above this, a three-position toggle switch, mounted for horizontal operation, controlled the flashing indicators, without automatic self-cancelling.

To the left of the ignition/lighting switch was a push/pull off/on switch for the single-speed, self-parking windscreen wipers. If optional windscreen washers were fitted, the pump control was to the left of this again. To the right of the ignition/lighting

Original rubber matting and the pedal arrangement. The black blanking rubber below and to the left of the dipswitch is an access point to the clutch slave cylinder bleed nipple. The hole in the matting below the pedals aligns with the footwell drain hole. Yes, BMC expected its sports car to fill up with water!

26

SPRITE MkI

Perhaps the only non-sporty thing about the Sprite? The over-large black plastic steering wheel, which so many owners soon discarded for a more appropriate item. Today, the original wheel is both a rare and highly-prized item – but still not at all sporty! This is a totally original dashboard with the correct optional windscreen washer control fitted.

The rev counter was a listed option, but very few purchasers in practice could possibly buy a new sports car without such an item. The few who did not were faced with a stark blanking piece containing two red main beam warning lights.

switch would be the optional heater control, which could be pulled outwards to stop air reaching the heater matrix. When pushed in, allowing heated air to enter the cockpit, the control could be rotated clockwise to start the single-speed blower. The choke control was mounted at top left, with the pull starter control in a matching position at top right. The functions of the choke, starter, wiper and heater controls were initialled in white on their black plastic knobs. If a fresh air unit was fitted, the knob would be marked with an A. The dipswitch was floor-mounted, to the left of the clutch pedal on both right- and left-hand drive cars. The horn was operated by a push button in the centre of the steering wheel boss.

A chrome-plated grab handle was fitted to the outer edge of the facia panel for the use of a nervous passenger! Below this handle, the facia panel pressing was perforated to allow the fitting of a radio. This section could easily be removed and the covering trimmed to suit. Also in this area, but on the reverse side on the panel, can often be found yellow crayon markings giving the trimmer the colour of the covering to be used on the panel during assembly, the calibration of the speedometer, and whether a heater control was to be fitted. Also to be found on the rear of the facia was a wooden block to act as a very large washer or load-spreading member for the grab handle. One could therefore (with tongue in cheek) say that a Sprite dashboard contained a real wood finish – even if it was out of view on the reverse side.

Placed centrally above the dashboard, and secured to the scuttle panel by two screws, was the Eversure rear-view mirror. It was adjustable via the ball end of its chrome-plated pedestal, the clamp around the ball end also serving to secure the mirror to its frame and backing piece.

ENGINE

In October 1951, Austin introduced the first of a new breed of small capacity engine, the A series, for the new A30. The four in-line cylinders contained a bore of 2.28in (58mm) and a stroke of 3.00in (76mm) to give a total swept volume of 803cc. With two overhead valves per cylinder, a compression ratio of 7.2:1 and a single downdraught Zenith 26 carburettor, power output was 28bhp at 4800rpm. In this form, the three main bearing crankshaft earned a reputation for fairly rapid bearing wear, a problem exacerbated by the use of a by-pass oil filtration system.

For 1956, the engine was enlarged to 948cc by increasing the bore size to 2.487in (62.94mm). Power output rose to 34bhp with an 8.3:1 compression ratio. The by-pass oil filtration system was replaced by a full-flow, replaceable element oil filter which, along with numerous other modifications, made the second generation of A-series engine a most reliable and efficient unit. In fact, very few components were interchangeable between the two capacity sizes, although externally the two engines looked very similar apart from the larger version having a new oil filter canister which hung vertically from an adaptor on the right-hand side of the cylinder block.

These mechanical improvements, along with the arrival of an improved four-speed gearbox, allowed the A30 to become the A35, the basis of which was incorporated into two new cars for 1958 – the Austin A40 and the Austin-Healey Sprite. Although the Austin A40 Farina was to continue with the A-series engine in exactly the same form as that fitted to the A35, a few simple modifications were required for the Sprite. Ironically, they were developed by the Morris Engine Division at Coventry which by now was producing A-series engines as part of the corporate plan.

These modifications included heavier duty crankshaft bearings, a stronger clutch, a new alloy inlet manifold to suit twin SU semi-downdraught carburettors, a separate exhaust manifold and a rocker cover with a re-positioned breather pipe stub. The vertical stub was now bent over to the horizontal and angled forwards to suit connection with the front carburettor air filter via a rubber hose. In order to ease serviceability of the oil filter element, the oil filter adaptor was changed to move the filter casing further away from the cylinder block and to rotate it through a compound angle, outwards and rearwards; the close proximity of the main frame rail beneath the filter would not suit the vertical A35 arrangement. A new pipe between cylinder block and filter adaptor was also required.

Power output now rose to 42.5bhp at 5200rpm with peak torque of 52lb ft developed at 3300rpm. The compression ratio remained at 8.3:1 – but no low compression alternative was offered for the Sprite.

To identify the engine in this form, a new prefix of 9C was carried on an aluminium plate riveted to the front right-hand side of the block top face. This was followed by the letter U to identify that a floor-mounted centre change gear lever was fitted, the last prefix letter being H for high compression. Engine numbers ran from 101 to 49201.

Structurally, the engine had a cast iron cylinder block with a detachable cast iron cylinder head secured by nine ⅜in diameter studs. No separate cylinder liners were normally fitted to the block, although these were available as a service item where the usual over-bore sizes could not be accommodated within the casting thickness. Internal water passages allowed for the circulation of cooling water throughout block and head with outlets on the top face of the cylinder head. The front outlet was fitted with the thermostat and outlet housing, with a smaller outlet at the rear of the head to serve the heater when one was fitted. Water entered the cylinder block through the front face via the bolted-on water pump.

The engine compartment of Bob Reid's low-mileage Sprite displays many original features. Hose clips would have originally been of wire rather than jubilee type, and the radiator cap is also a modern replacement.

The forged crankshaft ran in three main bearings of 1.750in (44.46mm) diameter. Each bearing had a removable cap located and secured by dowel bushes and bolts. The centre main bearing carried the four half-circle thrust washers controlling end float of the crankshaft and providing support against the thrust created by disengaging the clutch. Replaceable main and big end bearing shells were of the thin-wall, steel-backed type with a white metal bearing surface. Pressure-fed lubricating oil entered the crankshaft through holes in the main bearings to pass through internal drillings to reach the big end bearings.

The crankshaft drove an 8¼in (210mm) diameter flywheel, secured by a dowel and four bolts to a flanged face. The outer diameter of the flywheel carried a shrunk-on ring gear to engage with the inertia-type starter motor pinion gear. At the front of the crank, a sprocket drove the single row timing chain, the sprocket being located to the crank by a Woodruff key which also engaged with a pressed steel single vee pulley carrying a belt to drive the water pump, fan and dynamo. The timing gear drive was enclosed in a pressed steel casing with the cam sprocket containing two synthetic rubber buffer rings to act as a tensioning device for the timing chain. This was not an altogether successful idea as these rings could not cope with much stretch in the timing chain, resulting in a common A-series rattle in high-mileage engines. The camshaft ran in three bearings along the left-hand side of the engine, only the front bearing being a replaceable copper-lead linered type, the remainder running direct in the cylinder block. Each bearing was pressure-fed from the main lubrication system. The cam provided a valve lift of 0.28in (7.14mm) and was timed with the inlet valves opening at 5° BTDC and closing at 45° ABDC, and the exhaust valves opening at 40° BBDC and closing at 10° ATDC.

In addition to operating the valve gear, the camshaft performed three other duties. Towards the front, an additional eccentric drove the mechanical petrol pump located towards the front left-hand side of the block. A skew gear near the centre of the shaft drove an intermediate shaft located at right angles to the camshaft but inclined upwards towards the right-hand side of the engine. This shaft drove the distributor via an offset dog slot. The rear of the camshaft contained a peg drive for the oil pump located directly behind the cam in the rear of the cylinder block. A steel back plate bolted to the rear of the cylinder block with a separate cover to shield the protruding oil pump. This back plate provided an interface between the engine and bellhousing, and also located the Lucas starter motor on the right-hand side.

A steel front plate was also fitted between the front of the cylinder block casting and the timing gear cover. This plate extended outwards to provide location of the engine mounting rubber blocks either side.

Below the block, a flat-bottomed pressed steel sump contained 5.5 pints (3.7 litres) of oil with a single vertical baffle plate to prevent oil surge away from the oil pick-up strainer in the rear section of the sump. Lubricating oil was picked up through the submerged strainer gauze and passed through a suction pipe into a vertical drilling within the cylinder block casting. This connected with a horizontal drilling to the oil pump situated in a recess in the rear of the block. The pump pressurised oil to a maximum of 60lb/sq in and delivered it to the oil filter housing via the external pipe. Adjacent to the oil pipe union, a pressure relief valve controlled the maximum oil pressure, returning excess flow to the sump through an internal drilling.

Replaceable full-flow oil filter elements were of either paper or cotton and held in position in the casing by a spring-loaded plate. This plate could some-

Without the optional heater, accessibility to the battery, distributor and combined brake and clutch master cylinder is quite easy. The master cylinder here is a replacement unit with a plastic rather than steel filler cap.

With the optional Smiths heater (right), air is taken from the right of the grille through the long ducting pipe to the blower casing. A flap valve between blower and heater box controlled air flow by means of a dash-mounted knob. Heated air entered the cockpit via the plenum chamber beneath the heater box.

times become stuck to the bottom of the filter element and accidently be discarded when a new element was fitted. Without this plate to hold the filter into the correct position against the adaptor head, unfiltered oil was free to pass into the main oil passages. The adaptor head contained a by-pass valve which would open if a clogged oil filter failed to allow a sufficient passage of clean oil. A pressure differential of any more than 15 to 20lb/sq in between the inlet and outlet side of the filter would cause the valve to open and maintain lubrication to prevent seizure, but this would again be unfiltered oil and detrimental to good bearing life.

The oil filter head and bowl assemblies were supplied by either Tecalemit or Purolator. Externally, there was a visible difference between the two types. The raised dome of the Tecalemit filter head contained the threads into which the central filter bowl securing bolt was screwed. The Purolator filter head had only a very small raised filter head, for the centre bolt was shorter and screwed into a threaded portion which extended downwards from the head into the centre of the element.

Internal drillings within the cylinder block and head allowed all components to be adequately lubricated and cooled as required, all oil then returning to the sump under gravity for reuse. Due to the positioning of the oil pump above the sump oil level, initial start-up of a freshly assembled engine required the pump to be primed, so a priming plug was provided in the horizontal drilling on the suction side. This hexagonal plug was situated at the rear left-hand side of the engine just in front of the water jacket drain tap; when the plug was removed, a quantity of oil could be pumped directly into the drilling to fill the pump.

Two types of oil pump were fitted during production but cannot be identified externally, the rear back plate being the only means of access. The pumps were either a Burman eccentric vane type or a Hobourn-Eaton concentric type. In place of the low oil pressure warning sensor fitted to other A-series engines, a threaded union on the right-hand rear corner of the block (adjacent to the oil pressure relief valve) allowed the connection of a capillary tube leading to the oil pressure gauge.

The right-hand side of the engine carried the electrical equipment: the starter motor, distributor, dynamo, coil (this was relocated from the top of the dynamo to the body after engine number 9C-U-H 11888) and sparking plugs. This close grouping allowed a neat and simple wiring loom to be used.

The left-hand side of the engine carried the induction, fuel and exhaust systems, thereby keeping fuel and electrical components safely separated. Also on the left-hand side were two removable pressed steel tappet chest covers, which allowed access to the bucket type tappet blocks should a change of camshaft be required without having to remove the engine. However, the tendency for these covers to leak oil far outweighed any practical benefit.

The four aluminium alloy pistons with dished crowns were each fitted with three compression rings and one oil control ring. They were available in overbore sizes of +.020in, +.040in and +.060in. The gudgeon pins were a floating fit in each piston and secured to the connecting rod by a pinch bolt. The rods were offset between small and big ends to allow for the offsetting of the bores in relation to the crankpins. The big end caps were split at an angle to the vertical axis of the rod to allow these to be passed down the small diameter cylinder bores on assembly.

The cast iron cylinder head contained two vertically placed valves per cylinder, the inlets being of 1 3/32in (27.8mm) head face diameter and the exhausts of 1.00in (25.4mm). A single valve spring for each valve was secured by a cap and split collets, with a spring clip to retain the collet halves together above the valve cap.

The valves were driven by overhead rockers pivoting about a single hollow rocker shaft carried in four pedestals. The pedestals were secured to the cylinder head with studs, and in addition four of the head studs from the cylinder block passed through both the head casting and pedestals. Early engines used forged steel rocker arms with replaceable bronze bush bearings. From May 1959, pressed steel rockers began to be fitted, although forged rockers also continued for a while. The pressed steel rockers were made up of two halves, spot-welded together. According to the manufacturer, the bush in the pressed steel rocker was non-replaceable. The valve end of the rocker carried a hardened foot acting directly upon the valve stem. The opposite end carried a threaded adjuster and lock-nut to set the clearance between the valve stem and foot, the end of the adjuster being ball-shaped to fit into the socket end of the pushrod.

Each valve ran in a pressed-in cast iron valve guide and seated directly into the head casting. No valve seat inserts were fitted but were listed as a service item

A close look at the blower and heater unit with the cylinder head tap to control the flow of hot water. Four spring clips held the heater box together, usually one on top, one on the right and two on the left.

SPRITE MK I

With a heater fitted, this ducting panel channeled air from behind the grille into the flexible ducting pipe to the blower unit. This Sprite also has the Healey Motor Co Ltd anti-roll kit fitted.

Without the optional heater, the ducting panel alongside the radiator was not fitted. Note the additional bonnet support stowed across the front of the radiator, and the optional windscreen washer with the bottle correctly fitted to the inner wing support bracket.

to repair a damaged cylinder head seating. Valve stem oil seals were fitted to both inlet and exhaust valves, but from engine number 9C-U-H 1397 the seals and valves were altered as a production improvement. The valve cap was also changed to suit this modification.

Lubricating oil from the cylinder block passed through drillings in the head and front rocker pedestal to fill the hollow rocker shaft. Drillings in the shaft allowed oil to reach each rocker, with returning oil passing down the pushrod holes to lubricate the tappet blocks and cam lobes before reaching the sump.

The combustion chambers were of cardoidal (heart) shape, following a theory developed by Harry Weslake to promote very high combustion efficiency and good fuel economy. Austin, then BMC, were obliged to include a Weslake Patents number plate on all the engines they produced using this type of combustion chamber. This riveted plate was carried on the left-hand top face of the pressed steel rocker cover with a matching riveted 'Austin' plate on the right-hand side.

Inlet and exhaust manifolds were both fitted to the left-hand side of the cylinder head with six studs. Two inlet ports split off to feed each cylinder while a siamesed centre exhaust port served the centre two cylinders, the outer cylinders having individual ports. The room taken by the eight pushrod holes on the same side of the cylinder head necessitated this five-port arrangement.

The rocker cover was secured by two blind-headed nuts to two extended rocker pedestal studs, a cork gasket fitting between the cover and machined seating face on the head. An oil filler cap was provided at the front of the cover with a list of approved lubricants pressed into its metal cap. The cap was a bayonet fitting with a retaining wire incorporated to prevent it from dropping into the darker regions of the engine compartment. Engine ventilation was by means of a pipe from the rocker cover to the front air filter, while a steel tube provided an additional venting pipe from the front tappet chest cover to the atmosphere.

All engines were painted 'Engine Green', which could be described as a grassy green.

Only minor modifications were made to the 9C engine during production and these are listed in the 'Production Changes' table. However, both BMC's Special Tuning Department and the Donald Healey Motor Co Ltd developed and offered a wide range of tuning and competition options for this engine, described in the 'Options & Accessories' section.

BMC operated an exchange engine scheme, the gold-painted factory replacement units being known as Gold Seal engines. A different prefix was used: the Gold Seal prefix for a 9C engine would be 8G 10.

COOLING SYSTEM

A vertical flow radiator was bolted to the upright pillars immediately in front of the engine. Due to the low bonnet line, the bayonet fitting filler/pressure cap was placed in an extension neck protruding rearwards from the left-hand side of the radiator top tank.

The bottom tank contained a drain tap at its lowest point with the water outlet stub being on the left-hand side, water returning to the radiator through a stub on the right hand side of the top tank. A threaded boss on the right-hand end of the top tank housed the temperature sensing bulb for the water temperature gauge capillary tube.

A curved ducting plate was fitted to the rear of the radiator to shroud the fan blades. From chassis number AN5-6889 (November 1958), this shrouding was increased to include vertical panels to encase the fan totally from above and from each side. Also at this time, the number of gills increased from 10 to 12 per inch. The black-painted radiator was supplied by Serck Marston or Morris Radiators and fitted with a 7lb/sq in pressure cap.

The belt-driven fan at first comprised a single steel pressing to form two blades, but from engine number

9C-U-H 359 a second pressing was added to form a four-blade fan. The fan blades and the belt-driven hub were painted red on very early cars, but changed to yellow at an undefined point. The hub also drove the centrifugal water pump mounted to the front face of the cylinder block. The pump had a cast iron body with a screwed plug in the right-hand side to provide a lubrication point for the spindle bearings. Care had to be taken not to over-lubricate the bearings as the efficiency of the carbon sealing ring could be impaired.

Made by Concentric Manufacturing Co Ltd, the pump had an inlet stub from the bottom hose on the left-hand side, with a smaller vertical stub off this to correspond with a similar stub in the cylinder head above. A short hose connected these to act as a thermostat by-pass, ensuring a measure of water circulation when the thermostat was closed. This feature has proved to be a source of irritation on A-series engines, as it is a dreadful fiddle to replace this hose in the confined space between the water pump and cylinder head. A special thin-walled convoluted hose was developed to ease the task without having to remove either the water pump or cylinder head, but in practice this type of hose tends not to last as long as a conventional thick-walled water hose.

Water returned to the radiator from the outlet casting on the top front face of the cylinder head. The aluminium casting, secured by three studs, also provided access to the thermostat located in the top face of the cylinder head. A short right-angled top hose connected the outlet to the radiator.

The cylinder block had a drain tap situated on the left-hand side at the rear. The tap was changed from engine number 9C-U-H 6359 (October 1958) to provide a British Standard pipe thread instead of the previous taper thread. A fibre washer had to be used with this later tap.

The optional Smiths heater system comprised a black steel box containing the heater matrix, a blower fan unit in a plastic casing, an air valve and trunking. The heater box was fitted to the platform directly in front of the battery and drew hot water from a tap on the top rear face of the cylinder head. The flow of hot water could not be controlled from within the car. Water returned to a junction in the bottom hose through a copper pipe attached to the inlet manifold. If a heater was not fitted, a blank was fitted to the cylinder head in place of the tap and the bottom hose lacked the junction pipe.

Cold air was ducted from behind the radiator grille adjacent to the radiator. Trunking of 4in (102mm) diameter passed along the right-hand side of the engine compartment to the blower unit mounted on the top footwell panel. Air was then directed into the heater box through an air control valve placed between the blower and heater box. A cable controlled the valve positioning from a dashboard-mounted push-pull knob. Air would flow when the knob was pushed in, and then the single-speed blower fan could be operated by rotating the knob clockwise.

Hot air was directed downwards from the heater box into a plenum chamber below the platform. From here, it could be directed either into the footwells through letterbox flaps or onto the windscreen through ducting behind the dashboard. The black-painted, spring-loaded flaps were either open or shut, so final adjustment of the volume of airflow to screen or footwells was not possible.

If a heater was not specified, the platform duct would be fitted with a blanking plate, the dashboard control omitted, the ducting from the plenum chamber to screen ducts omitted, and the plenum chamber fitted with rubber bungs. The letterbox flaps, however, were always included.

The cooling system contained 10 pints (5.7 litres) of water plus a further ½ pint (0.25 litres) if a heater was fitted. The dashboard temperature gauge would normally record a working temperature of between 180 and 195°F, but this could rise in very hot weather or with arduous driving.

An option for countries with a hot climate was the Fresh Air system – basically a heater system without a blower unit or heater matrix. The black box mounted to the battery platform was smaller than that of the heater box and served to direct air from the flexible ducting pipe into the plenum chamber.

All hose clips were wire type rather than the more common replacement of jubilee clips.

From the very beginning, high water temperature caused concern and was addressed by a series of modifications in production, all of which became retrospective modifications to earlier cars where this problem was experienced. Apart from the modifications already mentioned, a copper-gilled radiator core was introduced at chassis number AN5-641 (March 1958), the thermostat changed from 75° to 68°C at engine number 1073 (May 1958), an improved cylinder head gasket with ferrules around the end water holes arrived at engine number 1169 (May 1958),

Twin 1⅛ SU H1 carburettors with the correct Cooper air filters. The damper retaining nuts were brass on the 'Frogeye' only. The letter 'O' stamped into the nuts would indicate that modified damper pistons are fitted.

and an improved cylinder head gasket with better water passage ferrule seals around all water holes was introduced at engine number 6733 (November 1958). Failing these measures, BMC also offered a thermostat blanking sleeve to replace the thermostat, allowing a greater flow of water around the system but providing some measure of restriction. Running an A-series engine without either a thermostat or blanking sleeve results in the rearmost cylinders running hot, too great a water flow at the front of the engine causing a 'dead' movement of water at the rear and localised overheating.

CARBURETTORS & FUEL SYSTEM

A departure from the A35's single Zenith VME 26 downdraught carburettor to two SU H1 instruments was essential for three main reasons. First, there was a lack of bonnet height clearance for a downdraught carburettor. Second, more power was required than could be obtained with the Zenith. Third, a true sports car just had to have more than one carburettor in the eyes of the buying public!

The SU carburettors chosen were the H1 semi-downdraught type of 1⅛in choke diameter. These were fitted to a cast alloy inlet manifold connected by a balance passage cast integrally with the manifold. A sheet steel heat shield was employed to protect the carburettors from rising heat from the exhaust manifold. From engine number 9C-U-H 17365 (May 1959) this shield also extended down to the AC Sphinx Y Type mechanical fuel pump mounted to the cylinder block below the exhaust manifold. Between each carburettor and the heat shield, a ⅝in (15.6mm) thick phenolic insulator block further prevented heat transfer from the engine.

The pump drew fuel through a ¼in (6.25mm) diameter pipe from the rear-mounted tank. The pipe was coiled just below the pump inlet to absorb engine movement. From chassis number AN5-41016 (June 1960) the coiled partition of the pipe was replaced by a flexible section. The pump could be primed by hand with a spring-loaded priming lever, the pump's maximum flow rate being 40 pints (22.8 litres) per hour.

The carburettor float chambers were placed to the outside of each carburettor, with the linkages for throttle and choke acting upon an interconnecting linkage between the carburettor bodies. Both throttle and choke were cable controlled.

Each carburettor contained a throttle return spring to close the throttle butterfly valve. A further spring acted between the interconnecting throttle linkage and heat shield to act directly upon the throttle cable. An organ-type throttle pedal pivoted from the floor with a link and rod linkage held in two alloy bearing blocks to transfer movement to the cable protruding through the upper footwell panel.

A rigid ¼in diameter fuel feed pipe rose from the fuel pump to the top of the manifold, where it divided into two and fed each float chamber via Smiths Petroflex piping. From engine number 1551, an overflow pipe from each float chamber directed excess fuel downwards, past the manifold.

Carburettor fuel metering needles were of types GG (standard), with alternatives of EB (rich) or MOW (weak). The jet size was 0.090in (2.29mm) regardless of the needle used. Fine adjustment of mixture strength could be made by raising or lowering the jet in relation to the tapered metering needle, by means of a brass nut beneath the carburettor body.

Movement of the metering needle was effected by the manifold vacuum acting upon the dashpot piston. The movement of the piston was further controlled by a hydraulic damper acting within the centre of the piston tube. Topping up of oil for the purposes was catered for by unscrewing the hexagonal brass nut from the top of each carburettor and removing the damper rod and piston.

From around July 1959 at an unspecified point, the damper pistons were reduced in length from 0.378in (9.6mm) to 0.308in (7.82 mm) to improve throttle response. The modified damper pistons could be recognised by the letter O stamped on the hexagonal brass nut.

Air filters were of a simple expanded gauze pancake type, one for each carburettor. The front filter was provided with a stub pipe to connect with the vent pipe from the rocker cover. Filters were painted crackle grey and carried a white and blue Cooper decal with the cleaning instructions.

The fuel tank, mounted beneath the boot floor, was retained by six ⁵⁄₁₆ UNF studs and nuts with rubber sandwiched between the tank and floor at the mountings. Holding 6 gallons (27.3 litres), this tank had a long filler neck which passed through the boot space and onto the rear shroud. The vented filler cap, painted body colour, was a twist fit to the filler neck. The tank and neck were painted black, and the fuel contents sender unit was mounted in a depression in the top face of the tank. An optional chrome-plated locking filler cap could be ordered.

EXHAUST SYSTEM

Made in cast iron, the three-branch exhaust manifold was the same as that used in the A35. However, on the A35, the top of the manifold provided a hot spot interface for this model's single Zenith carburettor manifold. As the Sprite had a completely different inlet manifold for its twin SU carburettors, the hot spot face was blanked over with a steel plate and gasket. No heat transfer path was therefore provided for the Sprite as exhaust and inlet manifolds were now totally separate.

ORIGINAL SPRITE & MIDGET

The 'domed' parking cap of the wiper motor fitted to 'nine-stud' windscreen Sprites. In this instance, Leaf Green paintwork was not a current colour for the 'nine-stud' Sprites.

The later type of wiper motor cap with a flat top. The black-painted panel is the blanking plate where the pedal box assembly would fit for left-hand drive. Bodyshells could easily be built with the steering wheel on either side because all holes and welded nuts were included in production.

A single 1¼in (32mm) downpipe mated to the manifold via a conical seating face and was secured by a two-piece clamp drawn together by two bolts. A support bracket was fixed between the bellhousing and exhaust pipe to absorb the torque reaction of the engine against the exhaust. Without this bracket, the joint between the downpipe and manifold would be strained, causing the joint to fail.

At the rear of the floorpan, a mounting hung from the heel panel to support the exhaust pipe as it passed under the rear axle. This single-piece pipe finished alongside the fuel tank, where the second and final joint was made into the straight-through, absorption type silencer. The silencer intake pipe sleeved over the main pipe and was secured with a clamp. A bracket on the top of the silencer body connected to a welded stud from the boot floor via an insulating rubber bush mounting. The silencer tailpipe exited at an angle towards the left-hand side of the car.

From chassis number AN5-5477 (October 1958), the exhaust pipe was modified to clear the revised splash plate in the engine compartment.

ELECTRICAL EQUIPMENT & LAMPS

The electrical equipment was of Lucas manufacture. A single 12-volt battery, designated type BT 7A for home market cars and BTZ 7A (dry-charged) for export cars, was of 43 amp/hr capacity and wired to positive earth.

The battery sat in a black Bakelite spill tray and was mounted on the central shelf above the clutch bellhousing. Access to the battery became restricted when the optional Smiths heater unit was fitted directly in front of this on the same shelf. The battery was retained by a black-painted clamp and two threaded rods.

The wiring loom was covered in black cotton with a blue fleck, all circuits being colour coded. Wiring connections were mostly of the 'bullet' type, while screw connections were used for the voltage control unit, fuse box, flashing indicator unit, dashboard switch gear, windscreen wiper motor, brake light switch, fuel tank sender unit and dipswitch.

The dynamo was type C39 PV2 with an extended armature spindle for the Smiths tachometer drive gearbox. The starter motor was M35 GI and contained four brushes. These two components were finished in BMC's familiar 'engine green'. The distributor was type DM2 PH4 with automatic advance and retard controlled by integral centrifugal bob weights and manifold vacuum via an external pipe from the front SU carburettor. The ignition coil, type LA 12, was originally mounted on top of the dynamo, but from engine number 9C-U-H 11889 (February 1959) it moved to a bracket welded to the corner gusset segment of the engine compartment, the revised location reducing vibration and heat. Low tension leads were screw fitted to threaded studs.

Spark plugs were 14mm Champion N5 unless a short-wave radio was fitted, when Champion XN8 plugs would be specified. Plain copper plug caps were fitted to black HT leads, but from 9C-U-H 4796 (September 1958) these were changed from an in-line fitting to a right-angled fitting, allowing the use of shorter HT leads. At 9C-U-H 25484 the caps were again changed to include suppressors, the caps now being black. Firing order, from the front, was 1-3-4-2. The HT leads were numbered to ensure correct fitment to each spark plug; prior to 9C-U-H 4796 the leads were fitted into numbered ferrules, but after this numbered short plastic sleeves were used. The leads were passed through an in-line plastic spacer just above the side-exit distributor cap.

Static ignition setting was 5° BTDC, this being set by lining up a V on the crankshaft pulley with a pointer on the bottom of the timing chain cover. A Micro adjuster was included on the distributor for small adjustments of ignition timing to suit individual conditions in service.

The voltage control unit mounted on the right-hand inner wing panel was type RB 106/1 or 106/2. The type SF6 fuse box was also mounted to the same panel and contained two 35amp fuses with space to carry two spare fuses. The cable-operated starter switch was mounted on a bracket on the heater/battery shelf above the bellhousing.

Headlamps were Lucas type F700, but the bulb specification varied according to the country of export, as shown in the table opposite.

Later European specification called for special headlight lenses of assymetrical beam pattern to meet local requirements. These changes for export versions necessitated a change of electrical socket within the headlight shell. Generally, the Lucas lamp units were of MkVI or MkX type, the latter having a stepped bowl housing and a much deeper lamp-retaining rim

SPRITE MKI

Country	Lucas number	Rating
UK RHD	414	50/40W
France LHD (from chassis AN5-7782)	411	45/40W
Europe LHD (excluding France)	370	45/40W
Sweden RHD (from chassis AN5-21118)	370	45/40W
Non-European LHD	415	50/40W

held by a spring to the bowl. From AN5-19015 (June 1959), North American Sprites used MkVIII or MkX units with sealed beam lights. MkVIII and MkX units also required a different body seating rubber gasket with extensions for the adjusting screws.

Control of dipped/main beam was by a floor-mounted switch alongside the clutch pedal. A red main beam warning light was incorporated in the lower edge of the rev counter or, if no rev counter was fitted, in the aperture blanking piece.

The front sidelight units, type 52338A, also acted as the flashing indicators by incorporating a twin filament bulb. The rear tail/brake light units, type 53330, were also used on other cars of the time such as the Morris Minor, MGA and Triumph TR2-3.

The number plate light unit contained one bulb until chassis number AN5-36194 (March 1960), whereupon a second bulb was added. Externally there was no difference in the two types, both of which carried chromed covers and were mounted directly above the number plate. On Australian-assembled cars, two number plate light units were fitted, one either side of the plate.

The windscreen wiper motor, type DR2, was mounted on the left-hand top footwell panel and drove two 8in (203mm) wiper blades by a rack and pinion drive. The single-speed motor allowed automatic parking of the blades in their lowest position. From chassis number AN5-5477 (October 1958), the shape of the motor self-parking cap/switch changed from a domed to a flat type, with the angle of wiper blade sweep reduced to 100° by the gearing.

The indicator flashing unit, type FL5, was screwed to the top right-hand edge of the bulkhead panel in the engine compartment. A single three-position switch mounted centrally on the dashboard panel controlled the indicators, a green warning lamp sitting between the speedometer and rev counter. The indicators did not self-cancel and gave little or no audible warning of operation because of the position of the flasher unit in the engine compartment.

A single horn was mounted to a bracket screwed to the base of the right-hand radiator pillar. An option of twin Windtone horns could be ordered, in which case these would be mounted to the left-hand radiator pillar as the heater air ducting location plate left insufficient room for them on the right-hand pillar. Twin horns were operated by a relay unit to protect the horn switch and wiring from their increased current requirements.

TRANSMISSION

The transmission was taken directly from the A35, very few alterations being made.

The Borg & Beck clutch was a single dry plate of 6¼in (16cm) diameter. Six pressure springs gave a total clamping force of between 540-600lb (245-272kg), an increase over the A35 figure. The release bearing was of special carbon graphite and operated by an hydraulic slave cylinder and pushrod, the system being self-adjusting in service. The alloy-bodied slave cylinder was changed to cast iron at AN5-6696 (October 1958) to improve wear resistance.

The gearbox was made up of three alloy casings containing the four forward speeds and one reverse. The main casing comprised the bellhousing and gear enclosure, and behind this the tailshaft casing housed the speedometer drive gears and rear main bearing and oil seals. Fitted directly above this was the gear change remote control casing.

The main casing had a smooth finish, lacking any external ribbing for strength. Access to the gear cluster was via a removable alloy plate on the right-hand side, an oil drain plug being provided at the bottom of the casing. The rear gearbox mounting was carried by the tailshaft casing and secured by four bolts to the car. Due to the enclosed design of the floorpan and transmission tunnel, it was not possible to remove or

The 9C-U-H engine and 'smooth casing' gearbox showing the correct oil filter mounting arrangement. This engine is fitted with larger 1¼in SU carburettors and ram pipes to one of the many tuning option specifications offered by BMC Special Tuning and the Healey Motor Co Ltd.

refit the gearbox with the engine in situ. On the left-hand side of the gearbox, a combined oil filler/level plug could only be reached by removing a large rectangular rubber bung in the side of the transmission tunnel, the gearbox holding 2⅓ pints (1.3 litres) of engine grade oil. The chrome gear lever, considerably shorter than that of the A35, was still fitted with the same black plastic knob with the change pattern picked out in white. Up to gearbox number 11520, the gear knob was accompanied by a locknut on the gear lever, but this was replaced by a knob with a rubber bushed centre which removed the need for a locknut. From engine number 3995, the lever was cranked to move the knob 1⅛in (28mm) rearwards, to prevent the driver's knuckles hitting the dashboard when selecting first or third gears. A pressed steel domed cover was fitted to the top of the transmission tunnel to house the gear lever rubber gaiter, the cover being painted body colour.

Gear ratios were far from ideal for a sports car, first and second being very low. Quite rapid getaways could be made from standstill in second, but the large step between second and third taxed the torque of the 948cc engine. Second, third and fourth gears were provided with cone synchromesh, although the effectiveness of this suffers with time, especially on second.

Gear ratios were as follows: first, 3.627:1; second, 2.374:1; third, 1.412:1; fourth, 1.00:1; reverse, 4.664:1. This poor choice of ratios soon led to the option of a close-ratio competition set being made available from BMC Special Tuning and the Healey Motor Co, these being: first, 3.2:1; second, 1.916:1; third, 1.357:1; fourth, 1.00:1; reverse, 4.114:1. These ratios could be obtained by rebuilding the original gearbox with new components or by buying a completely new gearbox under part number 22A 162.

From the gearbox tailhousing, drive was taken by an almost fully enclosed propeller shaft to the rear axle. A Hardy Spicer universal joint was fitted to each end of the shaft to allow for angular movement, each joint being provided with a grease nipple for periodic lubrication. As the forward joint ran within the enclosed propshaft tunnel, a rubber bung in the left-hand side of the tunnel could be removed for access. The rear wheels would then have to be rotated until the nipple became visible. A rubber bung was also included in the front inside footwell panel to allow access to the clutch slave cylinder bleed nipple.

It can be seen that much of the service attention to the transmission can be performed from inside the car rather than underneath it.

REAR AXLE

Made by Rubery Owen, the fabricated steel rear axle casing was a modified version of that used in the A35. Welded brackets provided the attachment points for the quarter elliptic springs with an attachment point for the radius arms above the casing. These brackets also provided a means of securing both the bump stop rubber blocks and the rebound check straps. Inboard of each bracket, a forward-facing bracket provided the attachment point for the drop link of the Armstrong lever arm shock absorber. One further bracket was welded to the case, this being an attachment plate for the handbrake linkage located beneath the left-hand axle tube.

The outer ends of the axle casing were threaded, allowing the ball raced hubs to be secured with 1⅞ A/F nuts. The left-hand side had a left-hand thread, although early axles were produced with a right-hand thread. An oil seal was fitted to the inboard end of each hub to prevent oil leakage into the brake drums. To the outboard end of each hub, an O seal and paper gasket were sandwiched between the hub flange face and flange face of the half shaft. The four threaded wheel studs were spline-driven into the hub flange with corresponding clearance holes in the half shaft driving flange and brake drum. To prevent the half shaft from breaking its seal with the hub when a road wheel or brake drum was removed, a single ¼in UNF countersunk screw secured the half shaft flange to the hub. Despite this, some leakage of rear axle oil is not unknown when a brake drum is removed. The cast iron brake drums were similarly held by two further screws to the hub to minimise leakage of oil when the road wheel is removed.

The differential gears were carried in an aluminium casting which was fitted into the front of the axle casing and retained by eight ⁵⁄₁₆ UNF studs and nuts. The hypoid bevel gears, comprising a nine-tooth pinion and 38-tooth crown wheel, gave a ratio of 4.22:1. With standard size 5.20-13 tyres, this resulted in a road speed of 15.4mph per 1000rpm in top gear.

A combined oil filler/level plug was provided at the rear of the axle, just right of the centre line with a drain plug at the base of the casing. There was an additional filler/level plug in the right-hand side of the differential carrier nose casing. This was due to the inheritance of the carrier from other vehicles in the BMC family which had no plug in the axle casing. Axle ventilation was by way of a plastic tube and dust cap fitted to the top of the casing to the right-hand side of the differential. Oil capacity was 1¾ pints (1 litre) of SAE 90 grade.

Although the differential gears were reliable, half shaft breakage was not unheard of in these early axles. This problem became exacerbated as the engines grew in size and power in later years. Breakage usually occurred on the inboard end of the shaft where the splined portion engaged with the differential. Fortunately, both half shaft and differential gear removal were easily accomplished and some owners became very well practised at it. This half shaft weakness was to be resolved in due course.

The only modification to the axle during produc-

A35 twin leading shoe drum brakes were felt to be adequate for the Sprite's weight and performance. In this view, the washer in front of the shock absorber lever arm is an owner addition, and the cup washer for the Healey Motor Co Ltd anti-roll bar link is too large.

tion occurred at chassis number AN5-4333 (September 1958) when the lever arm shock absorbers were slightly repositioned, entailing the attachment brackets on the axle to be suitably modified. The axle casing and differential carrier were painted black, and the axle serial number was stamped into the left-hand axle tube.

FRONT SUSPENSION

Dating from the Austin A30 of 1951, the front suspension had a new lease of life in the Sprite, and in fact continued on all subsequent Sprites and Midgets until production ended in late 1979. A total production life of 28 years may suggest that the original design philosophy was exceptionally good, but as most A30/35 and Sprite/Midget owners will testify, it was not without problems! Austin A40 Farina owners were also treated to the same suspension.

A pressed steel lower wishbone was pivoted at its inboard end by two fulcrum bolts held in brackets welded to the chassis rail. Metalastic rubber bushes, two per bolt, provided insulation between the wishbone and car. Up to chassis number AN5-11125 (February 1959) the fulcrum bolts were locked by castellated nuts and split pins, but Nyloc self-locking nuts were used thereafter. At the outboard end, the wishbone was forked with a threaded bush into each fork, a threaded fulcrum pin running in these bushes with the centre portion passing through the bottom kingpin eye. A cotter pin was used to lock the kingpin to the fulcrum pin. As the wishbone moved under road conditions, radial movement in relation to the kingpin was accommodated by the threaded portion of the fulcrum pin running in the threaded bushes in each fork end. A grease nipple was provided for lubrication, the fulcrum pin being suitably drilled to allow grease to reach the threaded bearing surfaces.

The vertical kingpin passed through two phosphor bronze bushes in the stub axle, the lower bush being of a larger diameter. Both bushes were provided with their own grease nipples in the stub axle, and dust-excluding tubes were provided to keep grease in and muck out.

The forged stub axle was drilled and threaded to take the forged steel steering arm which extended forwards to provide a tapered seating bore for the track rod end ball joint shank. The steering arms were revised from chassis number AN5-4800 (September 1958), requiring longer bolts to be fitted to the forward holes.

The stub axle also located the brake back plate with four 5/16 UNF bolts. The front hubs contained non-adjustable, angular contact ball races with a centre distance piece to space the inner races. A castellated nut and split pin retained the hub and bearing assemblies to the stub axle. The bearings were pre-packed with high melting point grease with a lip type seal on the inboard end to prevent grease penetrating the brakes. The hubs were changed in October 1958, not defined by chassis number, to a stronger and improved pattern. The later hubs carried the part number 2A4348, the early hubs being numbered 2A4137. The stub axles were commonised to fit either side of the car.

At the top of the kingpin, a trunnion block housed an Oilite thrust washer to act as a bearing surface against the upward thrust of the stub axle. The trunnion block was provided with shims in order to set the vertical free play of the stub axle correctly. A castellated nut and split pin secured the trunnion block to the kingpin, but replacement kingpins were subsequently equipped with a Nyloc nut and no drilling for a locking split pin.

The top member of the front suspension was the Armstrong double-acting lever arm shock absorber, the alloy body being left unpainted. This was secured by three 3/8 UNF bolts to its mounting platform, the black-painted lever arm extending outwards to connect with the trunnion block via a fulcrum bolt running in two Metalastic rubber bushes in the block. A pinch bolt in the end of the arm locked the fulcrum bolt into its correct position. A castellated nut and split pin fitted to the forward end of the fulcrum bolt further secured the bolt to the lever arm, which was forward of the trunnion block.

The coil springs were made from 1/2in (12.7mm) diameter spring steel and were wound to give seven working coils. They were rated at 271lb/in (19kg/cm), with a free length of 9.4in (238.8mm). The base of the coil sat in a pressed steel spring pan which was bolted to the centre of the wishbone with four 5/16 UNF bolts. The spring pan was made as a separate item to the wishbone to aid the removal and refitting of the spring in service. The top of the spring was located in the underside of the crossmember where it rose up towards the shock absorber mounting platform. A conical rubber bump stop was fitted in the centre of the spring's top location to contact the raised portion of the spring pan on full suspension deflection. The rebound rubber stop was located immediately below the shock absorber lever arm, which had a widened section to correspond with the rebound rubber block. Both rubber stops were a push fit into the front crossmember.

Suspension geometry at normal unladen ride height was as follows: camber angle, 1°; castor angle, 3°; kingpin inclination, 6½°. All suspension components were painted black, although the steering arms were marked with a splash of yellow.

In service, the threaded lower fulcrum pin and the corresponding threaded bushes in the wishbone forks were prone to wear quite rapidly, especially if the regular lubrication schedule was neglected. Both the fulcrum pin and wishbone had to be replaced when wear occurred. Difficulty was often experienced in attempting to remove a worn fulcrum pin from the wishbone as the pin would inevitably seize to the kingpin eye, making the unscrewing of the pin no easy task.

The kingpin was also subject to wear, particularly in the lower stub axle bearing. Moisture found its way into the lower bearing where it would attack the steel kingpin, particularly in the absence of regular lubrication. A new kingpin and new phosphor bronze stub axle bearing bushes would then be required, the latter having to be line-reamed in situ with a special reamer.

From chassis number AN5-48338 (October 1960), the wishbones and spring pans were revised to increase stiffness. The wishbones now gained additional bracing in the form of welded-in gussets in the lower face, necessitating new spring pans with scalloped (rather than straight) sides.

An anti-roll bar kit was available from the Donald Healey Motor Co Ltd. In order to fit this, the front face of each wishbone required drilling to accept the mounting plates.

Rear Suspension

The rear axle was hung on the end of the two quarter elliptic leaf springs. Made by Bramber, these were 21in (525mm) long and of 1¾in (44mm) width. Each spring comprised 14 leaves plus a master leaf to incorporate the rubber bush location for the axle. Ten leaves were of ⅛in (3.18mm) thickness, the remaining five of 5/32in (3.97mm). The load rate was 97.8lb/in (1.13kg/m) but an alternative heavy duty spring became available with a load rate of 118lb/in (1.357kg/m), this spring being made up of 11 leaves of 3/16in (4.76mm) thickness.

Each spring was located at the front into a fabricated box in the rear bulkhead. A heavy gauge anchor plate welded to the floor was drilled with four holes to take the spring clamping bolts. The front two bolts passed through vertical holes in the spring leaves to screw into a saddle clamp sitting on top of the spring. An extension plate welded to the clamp acted as a means of positioning the saddle when assembling the spring into the location box, the front of the spring being otherwise inaccessible. A further saddle clamp retained the spring to the anchor plate just outside the location box and was therefore fully accessible. This second saddle clamp was later replaced by an inverted U bolt.

The effective length of each spring was the 15½in (394mm) distance behind the anchor plate to the axle location bush. Three clamping straps were used to keep the leaves together.

With a quarter elliptic spring acting below the centre line of the axle, an upper link in the form of a radius arm was fitted parallel to the spring. The radius arm, made up of two steel pressings welded together along the top and bottom centre lines, formed a hollow, rectangular-section tube with a Metalastic bush fitted to each end. The front fitted into a large U-shaped bracket bolted to the rear bulkhead and gusset plate above the spring. The rear of each arm connected to the mounting bracket welded to the axle casing. Both bracket and arm were painted black.

The rear shock absorbers were Armstrong double-acting piston type, again with an unpainted alloy body. Each acted through a black-painted lever arm and drop link to a welded bracket on the rear axle casing inboard of the spring and radius arm bracket. The shock absorbers were secured by two ⅜ UNF bolts to the inboard sides of the gusset plates with the

The later radius arm bracket, which is noticeably larger, incorporated the shock absorber retaining bolts and effectively strengthened the bodyshell in this highly stressed area. The shock absorber was slightly repositioned to allow this, with accompanying revisions to the link and axle bracket. The spring was now secured by a U bolt.

Rear suspension detailing prior to AN5-4333 (September 1958), showing the smaller radius arm bracket at the front of the rear wheelarch. The bridge clamp for the spring should be threaded to take shorter bolts than shown here, the nuts not being required.

lever arms extending rearwards, the drop links containing rubber bushed joints to link the arms to the axle brackets. The left-hand lever arm was marked with a splash of yellow and blue paint and the right with white paint.

From chassis number AN5-4333 (September 1958) and intermittently to 4507, the shock absorbers were slightly repositioned on the gusset plates in order for the securing bolts to help secure new, larger radius arms brackets as well. The shock absorbers were given shorter lever arms and longer drop links. This also necessitated moving the drop link bracket extending forward and downward of the rear axle. The first 200 Sprites to be modified in this way carried a mark of red paint on the radius arm brackets to identify the revised design.

Controlling the extremes of axle deflection, a rubber block bump stop was fitted to the top of each outboard axle bracket above the radius arm attachment. Each block was mounted on a platform face and retained by two long split pins passing through the block and platform side flanges. Limiting axle drop on full rebound, two check straps hung from the axle tunnel to attach to an inward facing stud protruding from each axle bracket.

STEERING

A rack and pinion unit sourced from the Morris Minor provided very high-geared steering of just 2⅓ turns from lock to lock for a nominal 30ft (7.62m) turning circle.

The rack unit was held in two alloy clamping brackets bolted to the front crossmember ahead of the wheel centre line. A spacing shim would be selected and fitted behind one of the brackets to align the steering column in relation to the top column mounting. A track rod extended from either side of the unit with the innermost ball joint contained within a convoluted rubber gaiter. The outer end of each rod was threaded to screw into a Lockheed ball joint, and the tracking angle of the front wheels was set by the amount of thread engaging into each ball joint. Tracking was normally set within zero to ⅛in (3.175mm) toe-in, and a lock nut was tightened against the end of the joint to prevent the track rod rotating and altering the tracking angle. From AN5-22583 (July 1959), a moulded nylon retaining washer was added to the pinion shaft to improve the effectiveness of the felt oil seal.

A tapered shank rising from each ball joint engaged with a correspondingly tapered hole in the front of each forged steel steering arm, the rear of the arm being secured by two bolts to the stub axle. A castellated nut and split pin locked the shanks into the steering arms. A grease nipple was provided beneath each ball joint for periodic lubrication. A nipple was also fitted to the rack and pinion unit to the opposite end to the steering input shaft. This however, required SAE90 grade oil, not grease as many owners have mistakenly administered. Stronger steering arms of thicker section were introduced from chassis number 4800 (September 1958), requiring longer bolts to be fitted to the forward-most holes.

The steering column connected to the rack input shaft via splines and a pinch bolt, the end of the column being split to allow for contraction by the pinch bolt. The lower part of the column was fully exposed in the engine compartment, but as the column passed through the bulkhead panel, through a convoluted rubber grommet, it became enclosed by a non-rotating outer column. The outer column was secured by a two-piece aluminium clamp and bracket by two bolts to the bulkhead panel directly behind the dashboard. Spacer shims were included as required between the bracket and bulkhead to align the column's vertical angle properly. An oil-soaked felt bush acted as a top bearing between the inner and outer columns. The upper portion of the inner column passed through the dashboard panel to meet the steering wheel with a spline and taper seating.

The somewhat over-large 16in (406mm) steering wheel had a black plastic covering and two horizontal spokes with a ribbed pattern finish. The centre boss of the wheel contained the horn push button with a 'lightning flash' motif in chrome on a black backing.

All steering components were painted black with the exception of the aluminium brackets, which were left in a natural finish.

BRAKES

Lockheed hydraulically operated drum brakes were fitted to each wheel, the four 7in (177.8mm) cast iron brake drums each having two brake shoes of 6.75in (171.45mm) diameter × 1¼in (31.75mm) width, giving a total frictional area of 67.5sq in (435.48sq cm).

The front brakes contained two slave cylinders per side to provide a twin leading shoe arrangement, while the rear brakes contained a single cylinder each to provide a single leading and single trailing shoe arrangement.

Adjustment of each front brake shoe was by a snail cam 'Micram' adjuster placed between each shoe and its slave cylinder. Adjustment could be made with a screwdriver through an access hole in both the brake drum and road wheel. The rear brakes followed a similar means of adjustment, but with only one snail cam adjuster placed on top of the slave cylinder, the cylinder was free to move in a slot to equalise the pressure applied to each of the two brakes shoes. It was, therefore, quite a simple job to adjust each brake by raising the wheel clear of the ground, removing the chrome hub cap, pulling out the rubber dust excluder bung from the access hole in the drum, and

Original Sprite & Midget

Healeys offered a disc brake conversion along with knock-on wire wheels. Also fitted to this Sprite is the Healey anti-roll bar kit. Both of these features were introduced by BMC in later years, but differed from the Healey specification seen here.

inserting a screwdriver.

The handbrake operated on both rear wheels through a cable from the transmission tunnel to the rear axle. Here a pivoting compensator transferred movement to each rear brake via a rod to each. A grease nipple was provided on both the outer cable and compensator.

Due to the assymmetrical shaping of the transmission tunnel, the chrome-plated handbrake lever could only be accommodated on the left-hand side of the tunnel for *both* left- and right-hand drive cars. The lever, of conventional button release but not a 'fly-off' type, was secured to the tunnel by a black-painted steel bracket with the cable being operated from within the tunnel. Adjustment of the cable was by means of a screw-threaded portion on the very end of the outer cable. Two nuts could be rotated to alter the length of cable entering the compensator bracket on the rear axle.

The foot brake lever operated through a short adjustable pushrod to the master cylinder set in the pedal box assembly above the driver's feet. The pedal box, a sheet steel fabrication, contained both brake and clutch pedals. The single cast iron master cylinder contained two bores of ⅞in (22.2mm) diameter arranged in parallel, one bore to serve the brakes and the other for the clutch, Lockheed Super Heavy Duty Brake Fluid being specified.

Pedals and pedal box were painted black, both pedals initially having a rubber tread with the Austin 'A' motif, although this was changed at an early stage for treads of a vertical ribbed pattern without the motif. A common reservoir of hydraulic fluid served the brake and clutch systems, with a single screwed metal filler cap provided in the removable alloy top plate of the cylinder assembly. From the master cylinder, a ³⁄₁₆in diameter pressure pipe led to a brass union body bolted to the right-hand inner front wing panel. Here, a pipe connected to each front brake, while a fourth pipe passed along the underside of the floorpan to a bracket on the rear heel panel. A flexible hose connected this pipe to a brass union bolted to the differential casing, and from here a separate pipe connected to each rear wheel cylinder via a further brass union which also contained the bleed nipple. A pressure-operated switch for the brake lights was accommodated in the four-way brass union body on the front inner wing panel. A flexible hose to each front brake served the rearmost slave cylinder and a steel transfer pipe, external to the brake backplate, completed the circuit to the forward slave cylinder.

The full disc brake and wire wheel kit from Healeys used Girling disc brakes, larger Riley One Point Five rear brakes, and separate master cylinders for brake and clutch in a new pedal box. Healeys later offered a Lockheed disc brake system which used the original combined master cylinder and did not require Riley rear brakes or wire wheels to be fitted.

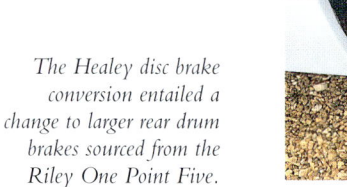

The Healey disc brake conversion entailed a change to larger rear drum brakes sourced from the Riley One Point Five.

None of the three flexible hoses were shrouded by coils to prevent chafing or damage, although a rubber sleeve was fitted to offer some protection.

Bleed valves were fitted to each brake, the front ones being in the forward cylinders only. No vacuum servo assistance was fitted or offered as an option, as pedal pressure was not considered to be excessive by the standards of the day. The four front slave cylinders were made of cast iron while the two rear cylinders were aluminium alloy; all used steel pistons.

The Healey Motor Co Ltd offered a disc brake conversion to replace the Lockheed front drum brakes, but this also entailed a change to wire wheels. Available as a complete kit of parts, the brakes were made by Girling under licence from Dunlop, the inventors of the disc brake. This conversion also entailed a change to 8in (203.2mm) rear brakes, and a new master cylinder and pedal box arrangement whereby the brake and clutch circuits were now served by separate master cylinders.

WHEELS & TYRES

Made by Rubery Owen, the road wheels, of 13in diameter and 3.5in rim width, were formed with a 'D' rim pattern. Secured by four ⅜ UNF studs to the hub, each wheel was made up of a pressed steel centre hub and a rolled steel rim. Up to chassis number AN5-15150 (April 1959) the two sections of the wheel were riveted together, but after this they were welded to improve their strength.

The wheel pattern was unique to the Sprites in this rim size, but closely matched that of the steel-wheeled MGA and MG Magnette. Twelve 'ventilation' holes of ¾in diameter were pierced into the disc face of the wheel in order to, so it was claimed, "improve cooling to the brake drums". Whether this did in fact perform such a task is a matter of debate, but it did make an otherwise plain wheel look rather attractive! To complement this, the chromed steel hub caps carried large depressions proclaiming the letters 'AH' for Austin Healey. Two further 1in diameter holes were pierced in the wheel hub face to correspond with a similar hole in each brake drum, allowing brake shoe adjustment without having to remove the wheel.

The wheels were again strengthened at chassis number AN5-39224 (May 1960), the hub face pressing having larger radiating ribs where it formed into the hub cap mating ring.

Wire wheels were never offered as an option by the factory, but the Healey Motor Co Ltd produced a complete wire wheel kit for the more serious owner to fit as an 'after-market' accessory. Dunlop Gold Seal 5.20-13 tyres were standard, these being a tubeless crossply of four-ply construction. Heavier duty Dunlop Fort six-ply tyres could be ordered for an additional £7.2.6d. Recommended inflation pressures were 18lb/sq in front and 20lb/sq in rear. The owner's handbook recommended increasing the pressures on the kerb side by a further 2-3lb/sq in, the benefits, it was claimed, being easier handling and improved tyre wear! The Schrader tyre valves were fitted with conical chromed dust caps, while the spare wheel cap was slotted for use as an inner valve remover. White-wall tyres became an option on North American market Sprites.

IDENTIFICATION, DATING & PRODUCTION

Rather confusingly for a car of monocoque construction, BMC issued both a body and a chassis number for the Sprite. For most purposes, the chassis rather than body number is regarded as the true identification number.

The chassis number appears on an aluminium plate

A complete set of tools, including the additional items offered by BMC. The two engine-lifting brackets were not a listed option for the Sprite tool kit, but were available from BMC dealers as a general item.

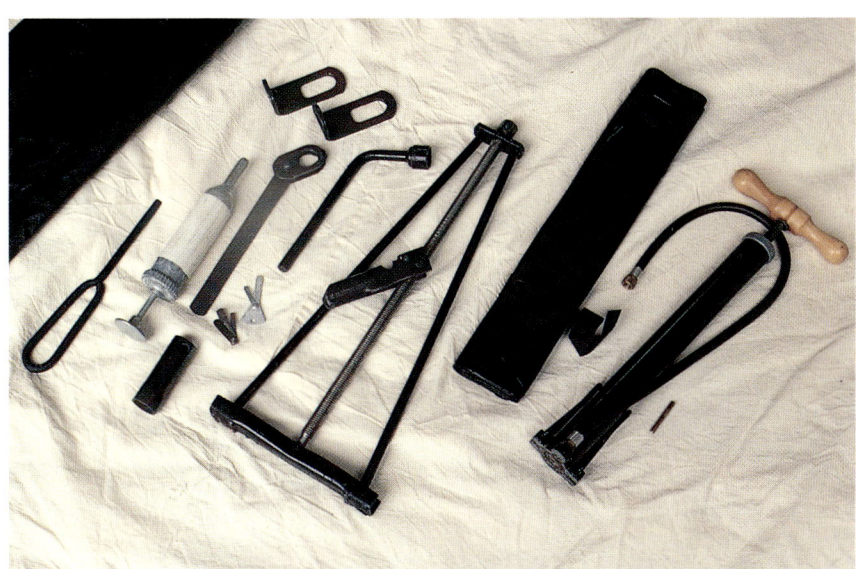

Ventilated steel wheels by Rubery Owen looked attractive but were suspect in the strength department. These were revised twice during production and were not used on any other vehicle, excepting the Innocenti. The chromed steel hub caps never had the 'AH' lettering painted.

ORIGINAL SPRITE & MIDGET

Under the Hardura trim covering the axle tunnel, the body number was crayoned to assist identification during the original assembly process. The underside of the bonnet would have been similarly marked.

The chassis (or car number) plate was stamped with the relevant number but never the engine number, only the wording 'see engine'. The plate was retained by two self-tapping screws. The number and wording should read from the left-hand side of the car, as here.

screwed to the sloping 'chassis' rail beneath the air filters. This carries the wording 'Austin, Made in England, When Ordering Replacements Quote'. Below would appear the actual number. This would be described as the 'Car Number' with a space for 'Engine Number' below that. In all cases, the engine number was never given on this plate, the wording instead reading 'See Engine'.

The chassis number prefix for the MkI Sprite was **AN5**. The **A** was the code letter for a small capacity (under 1200cc) A-series engine, the **N** was the BMC code for an open two seater sports car, and the **5** was the fifth series of Healey, the previous four numbers being allocated to the Austin-Healey 100 series. A further prefix letter **L** would denote a car built with left-hand drive.

The first production chassis number allocated was 501 and the final number 50116. Simply subtracting the first number from the last does not give the true number of Sprites built as some of the allocated numbers were never used on cars fully assembled at Abingdon, as many were despatched as Completely Knocked Down (CKD) kits to be assembled overseas. The largest market for CKD kits was Australia, where approximately 1000 Sprites were assembled by the Pressed Metal Corporation (a BMC subsidiary) of Enfield near Sydney from late 1959 to 1961. All Sprites assembled here were identified by **YH** in front of the chassis prefix. Following the prefix, a year code number would preceed the actual chassis number, with a local build number following on from this again.

Thus, an Australian-assembled CKD Sprite would follow this example: YHAN 5-9/44420/798. This represents: **Y** Australian assembled, **H** Healey (not used on MkI Sprites elsewhere), **AN5** as per Abingdon, **9** year code, **44420** chassis number in accordance with Abingdon-issued numbers, **798** the 798th CKD assembled. The practice of year coding was nearing an end as the Sprite was being assembled, so only the following code numbers would have applied: **8** 1 December 1959 to 30 November 1960; **9** 1 December 1960 to 30 November 1961. After 30 November 1961, the number was simply omitted.

This information was carried on a larger and re-positioned aluminium plate located on the bulkhead panel adjacent to the battery earth cable. The plate carried the 'Flying A' Austin logo with the warning that the guarantee ceased upon removal of the plate. Unlike the Abingdon-fitted plate, the engine number and body colour were also stamped into the respective spaces below the chassis number. The type of paint and paint manufacturer were also given in the form of prefix and suffix code numbers to the colour name. The prefix indicated paint types as follows: **1** Enamel, **3** Lacquer, **4** Acrylic Enamel. The suffix indicated paint manufacturer as follows: **1** BALM, **2** Lusteroid, **3** Bergers, **4** Brolite, **5** Taubman, **6** Spartan. It is not certain if all of these combinations were used although enamel appears to have been the usual paint finish.

The body number, commencing with the prefix BAE, was stamped on a plate screwed to the right-hand door hinge pillar of all cars, regardless of market. On Australian CKD cars only, this was also stamped into the top footwell panel near the chassis plate. To add further confusion, two more AN5-prefixed numbers may also be found on the bodyshell, but do not relate numerically to either the chassis or body number. The first is carried on a plate screwed to the left-hand door hinge pillar and is incorrectly referred to as the body number in the factory service parts manual. The second is stamped into the top footwell panel close to and above the master cylinder aperture on the right-hand side. These numbers do not appear in the factory final build records and were most likely used for internal factory purposes between the various manufacturing and assembly locations.

The body number on all cars could also be found hand-written in crayon on the inside of the bonnet panel and on the rear axle tunnel. This was to ensure that after the bonnet was removed from the body

This view of the right-hand door pillar shows the body number plate below the top hinge. This plate, with 'BAE' prefix, bears the proper body number. A similar plate in the left-hand pillar carries a number with an 'AN5' prefix, but this is neither the chassis nor body number.

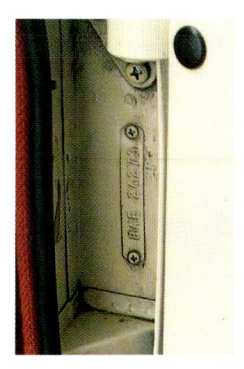

Stamped into the upper footwell panel is yet another number. This is not the actual chassis number, although it does approximate to the chassis number. This identification is thought to have been stamped by John Thompson Motor Pressings, who manufactured the basic monocoque tub. In this shot, the wiring has been released from its tab to expose the number.

Note to table
[1] LHD CKD totals exclude Innocenti kits.

upon arrival at Abingdon for the engine and gearbox to be fitted, it would re-unite with the correct body at a later stage, preventing any mismatch of panel fit or colour quality.

The engine number was carried in raised characters on an aluminium plate riveted to the top face of the cylinder block adjacent to the dynamo. The prefix was **9C-U-H**, **9C** being the Sprite MkI identification, **U** indicating a centre change lever gearbox, and **H** meaning high compression (but no low compression was available). Engines bound for Innocenti in Milan for use in the Sport 950 Spyder carried a 9CA-U-H prefix and were numbered from 101 to 750. Although it used no bodywork or panels from the Abingdon factory, Innocenti drew chassis numbers in accordance with the allocation for the AN5 series.

Due to assembly procedures, there was no unified numbering of chassis, body and engine numbers. The first chassis number did not necessarily contain the first body or engine number. However, the factory did record each chassis number and the number of body and engine apertaining to it.

Both gearbox and rear axle were numbered but without a prefix. The gearbox number was stamped into the top of the casing, and the rear axle number was stamped to the front of the left-hand axle tube.

Factory records containing the relevant numbers, specification, build and despatch dates are kept by the British Motor Industry Heritage Trust, Heritage Motor Centre, Banbury Road, Gaydon, Warwick CV35 0BJ. For a fee, a Production Record Trace certificate containing the factory build information can be issued to substantiate a car's original statistics. This service is available for all Sprites and Midgets built at Abingdon and Cowley.

PRODUCTION FIGURES

Year	RHD Home	RHD Export	LHD Export	RHD CKD	LHD CKD[1]	Total
1958	1927	706	5898	169	29	8729
1959	4042	396	16908	220	0	21566
1960	3972	339	13441	872	24	18648
1961	0	0	0	44	0	44
Total	9941	1441	36247	1305	53	48987

PRODUCTION DATING

Date	Chassis number	Notes
Mar 58	501	First production Sprite on 31 Mar
Jan 59	8927	First car built in 1959
Jan 60	30215	First car built in 1960
Nov 60	49584	Last fully-assembled Sprite at Abingdon
Jan 61	49821	First car built in 1961 (CKD)
Feb 61	50116	Last CKD (Innocenti) on 16 Feb

OPTIONS, EXTRAS & ACCESSORIES

To offer the Sprite at an attractively low basic price, the level of equipment and fittings was by necessity rather modest. Most owners felt the need to improve on the basic specification and could order from a list of options and accessories through their supplying dealer. The list fell into three main categories: those fitted at the factory, those fitted by the supplying dealer, and those available from BMC's own Special Tuning Department. Further to these, Donald Healey Motor Co Ltd also offered its own range of optional equipment. By virtue of having its own BMC franchise to sell Sprites, a brand new Sprite could be ordered through Healey fitted with any of the company's additional items.

To look at the factory-fitted options first, these comprised the following (with part numbers):

Front bumper bar with overriders and number plate mounting Exported Sprites were fitted with this as standard. AHA/5144.
Laminated windscreen Nine-stud frame, 14A 4719; later two-stud frame, AHA 5317.
Smiths heater kit 8G 9023.
Fresh air kit 8G 9024.
Sliding sidescreens Standard with optional hardtop, standard on all cars from AN5-34558 (March 1960), AHA 5414/5.
Rev counter 2A 9068.
Windscreen washer kit (Tudor twin jet) Prior to AN5-26724, AHA 5220; from AN5-26724, AHA 5501.
Windscreen washer kit (Trafalgar single jet)
Tonneau cover (black) AHA 5256 (RHD) or AHA 5257 (LHD).
Tonneau cover (white) From AN5-13543, AHA 5482 (RHD) or AHA 5483 (LHD).
Hardtop (ivory white) From AN5-8283 (December 1958), AHA 5485.
Locking petrol cap 2A 504.
Six-ply tyres Dunlop Fort.
Whitewall tyres USA only.

Note: Australian-assembled Sprites were equipped with the following items as standard: front bumper, rev counter, windscreen washer, locking petrol cap and tonneau cover.

Quite probably, a purchaser not ordering any of the factory-fitted options at the time of placing an order could subsequently have these fitted by the supplying dealer. In addition, BMC dealerships could offer a further range of options from the Approved BMC Accessories list, as follows (with part numbers):
Twin Windtone horns With relay.
Luggage rack Only from AN5-4333 (September 1958), AHA 5467.

Original Sprite & Midget

Peter Seamen's Iris Blue Healey Sebring Sprite, showing the correct wire wheels and disc brakes with which all Sebrings would have been fitted. Additional equipment and the state of the engine tune would have been to a customer's requirements, so quite possibly no two Sebring Sprites were identical. The bung below the door pillar covers the jacking point. This, and the shape of the sill, were unchanged throughout production of all Sprites and Midgets.

Radiomobile radio Models 20X (medium/long wave), 22X (medium wave), 230R (short wave).
Fire extinguisher 5½×1½in pressurised jet, 97H752.
Fire extinguisher 9½×1½in pressurised jet, 97H753.
Tyre pressure gauge 3H1875.
Tyre foot pump 2H3570.
Inspection lamp 17H5487.
Reversing lamp 97H2150.
Wheel rim embellisher 7H9714.
Wing mirror (boomerang type) Convex glass, 7H9829.
Wing mirror (ratchet type) Convex glass, 27H9650.
Exhaust deflector With Austin motif, 97H467.
Electric clock 1G2892
Interior prismatic mirror 7H9774.
Ash tray With Austin motif, 17H9558.
Child's seat Fitting on transmission tunnel, in red (AFA730), blue (AFA731), black (AFA732) or green (AFA733).
Seat covers Made to order
Travelling rug All wool, available in eight tartans, 27H9635.
First aid kit AJA5098.
Car polishing mitt Lambswool, 97H698.
Exhaust pipe trim 27H9806.

From very early on **BMC Special Tuning** offered a range of equipment to improve the Sprite's performance and driveability, but with the strict understanding that using such equipment would affect the normal warranty. Although BMC was quick to point out that reliability might be compromised by increasing power output, it stated that general reliability should be no worse than other cars offering comparable performance.

A Special Tuning Booklet, part number AKD 1021, set out the specifications, which basically amounted to five stages of tune:

Tuning Condition 1 A lightly 'cleaned-up' cylinder head with attention to improving gas flow in the ports and matching of the inlet and exhaust manifolds to the cylinder head ports. Power rose by 2.5bhp.

Tuning Condition 2 This entailed the fitting of flat top pistons (part number 2A 946) to raise the compression to 9.3:1 in addition to the first tuning condition. With the ignition set to 3° BTDC, power output rose to 47bhp at 5500rpm.

Tuning Condition 3 Stages 1 and 2 plus the fitting of a new camshaft (part number 2A 948) along with a new distributor (2A 951) and heavier valve springs (2A 950). With GG or GM carburettor needles fitted, up to 50bhp at 5800rpm could be obtained.

Tuning Condition 4 The cylinder head inlet and exhaust ports now came in for further attention by

The Vynide-covered dashboard contained a perforated area that could be removed to fit a radio, the Vynide having to be cut around the resulting aperture. David Whyley's Sprite displays an original Radiomobile radio, windscreen washer and fire extinguisher from the BMC options list. The passenger's grab handle is an original feature to all MkI Sprites.

SPRITE MKI

Safety? Hmm, the BMC optional child's seat meets no known safety standards today. An interesting and very rare period piece in Dennis Acton's Sprite. Did the colour fade through fright, one wonders?

The Healey hardtop predated the BMC item, which shared the same styling. The Healey hardtop, however, used deeper sidescreens, as shown here. Each hardtop was numbered, and identified by a plate in the right-hand door window edge. The highest number known is 190, but there may have been many more.

enlarging and reprofiling the tracts according to the dimensions given in the booklet. A further 2bhp could be obtained, but with a warning that this work needed to be carried out carefully to avoid damage to the cylinder head.

Tuning Condition 5 The fifth stage comprised a special exhaust manifold (AHA 5448) and exhaust system (AHA 5449 pipe and ARA 98 silencer) to increase output to 54/55bhp – or a 17.2% increase over the standard Sprite's output of 46.5bhp.

Further detailed improvements were also discussed in the booklet for the more serious owner to consider should even more power or stress be placed upon the engine. However, to complement the five conditions of engine tuning, the transmission could also be improved with a close-ratio gearbox (22A 162), a competition clutch assembly (C-AEJ 31) and alternative crown wheel and pinion ratios with matching speedometer, as follows:

Ratio	Part number	Speedometer
4.55:1	8G 7129	BHA 4118 (mph)
		BHA 4119 (kph)
3.73:1	ATA 7240	BHA 4120 (mph)
		BHA 4121 (kph)
3.9:1	C-ATA 7354	BHA 4122 (mph)
		BHA 4123 (kph)

Finally, to improve roadholding, stiffer front shock absorbers (C-AHA 6451 and 6452 for right-hand and left-hand) could be fitted along with adjustable-rate rear shock absorbers (C-AHA 6453 and 6454 for right-hand and left-hand), but only after the revision to the shock absorber mountings had occurred from chassis number AN5-4333.

The Donald Healey Motor Co Ltd of Warwick not only fulfilled its contractual obligation to BMC to develop the Sprite through research and a competition programme, but also created its own range of options and accessories. A subsidiary, Healey Speed Equipment Ltd of Grosvenor Street, London W1, was set up to market this equipment. The performance equipment was generally designed to exceed the level offered by BMC Special Tuning, and was therefore aimed very much at the serious owner who had competition or very fast road work in mind. Healey successes at major motor sport events throughout the world added to the catalogue of parts subsequently offered, perhaps the most famous of these successes leading to the term 'Sebring Sprite'.

The Sebring Sprite topped a list of options which offered a comprehensive package of parts, although it should not be confused with the special-bodied Sebring Sprites of John Sprinzel that were to follow. Sprinzel, a highly successful Sprite driver in international competition, established Healey Speed Equipment Ltd for the Donald Healey Motor Co Ltd. From this, he then went on to form his own company in Lancaster Mews, London and continued development, the culmination being his own Sebring Sprite. This racing derivative used a restyled fibreglass bonnet, a light alloy fixed-head hardtop which included the windscreen frame, a light alloy rear body section of normal Sprite profile, fibreglass door frames with light alloy skins, and special sidescreens.

Accepting that the name Sebring Sprite is used in conjunction with both Healey and Sprinzel, and not wishing to become too side-tracked into motor sport, I return to Healey's own listings for the Sprite.

Wire wheels and disc brake kit Five 60-spoke wire wheels, splined hubs and two-eared knock-on caps. Special 8½in (215.9mm) diameter front discs with Girling calipers. Larger 8in (203.2mm) diameter by 1½in (38.1mm) wide rear drums from the Riley One Point Five. Revised pedal box assembly (painted red) with separate brake and clutch master cylinders. In early 1961, Healey offered a new disc brake conversion using 8.75in (222mm) diameter discs with Lockheed calipers. This used the original type master cylinder although this required a simple modification to be made. More significantly, the new system did not demand the change to wire wheels. This system was termed the 'thin disc' at the time.

Engine Equipment Offered in parts and various stages, but with the addition of larger 1½in SU H4 carburettors and a rather neat 'Healey'-inscribed alloy rocker cover, which could also be obtained with a quick-action flip-top oil filler cap fitted towards the rear of the cover.

The ultimate offering was the Shorrock supercharger kit which boosted power to a claimed 68bhp at 5700rpm on an otherwise standard engine, or up to 85bhp with other performance parts fitted. The

Original Sprite & Midget

The optional BMC hardtop (far left) has visibly lower sidescreens than the otherwise similar Healey hardtop. As with the Healey version, each hardtop carried a number in the right-hand sidescreen window edge. A BMC hardtop (left) showing the over-centre clips that engaged with the slot in the screen top rail.

Shorrock C75B supercharger was an eccentric vane type fitted to the front left-hand side of the engine. A new crankshaft pulley, with two additional grooves, drove the supercharger via two vee belts. A single 1½in SU carburettor was mounted on the end of the supercharger intake pipe, and no air filter was fitted. Maximum delivery pressure was 7lb/sq in, controlled by a valve on the supercharger outlet pipe. Lubrication of the supercharger bearings was from a 'T' piece set into the oil pressure gauge take-off union via a small-bore pipe to the supercharger. Being a non-return system, lubricating oil was lost into the supercharger, becoming digested by the incoming fuel/air mix to be burned in the combustion chamber. Obviously, a supercharged engine would therefore consume oil at a higher rate.

Due to the higher positioning of the carburettor, it was found necessary to provide a small bulge in the bonnet to clear the dashpot top. In practice, although a considerable increase in both power and torque could be obtained, problems of overheating and blown head gaskets were likely to be experienced. The supercharger kit was phased out, possibly before the introduction of the Sprite MkII model.

Interior An attractive wood-rim steering wheel with three slotted alloy spokes and standard horn push. Lightweight competition seats moulded in fibreglass to fit existing seat mountings and available in the standard trim colours. An alloy boot rack was offered, although BMC also had its own item available. An anti-roll bar reduced the roll of the front suspension. There was an oil cooler too, plus many other items to, as the company slogan proclaimed, 'Make it a Spritelier Sprite'.

Hardtop Healey offered its own fibreglass hardtop, designed and made by Jensen. This required taller sidescreens than the BMC type. These were constructed with a steel frame covered in white or black Vynide, the panes being perspex. An aluminium-framed sidescreen of similar dimensions for use with the Healey hardtop also became available: this had a flat top edge as opposed to the slightly curved top edge of the Vynide-covered frame, thus producing a slightly mismatched gap with the top of the aperture.

The Donald Healey Motor Co Ltd wood-rim steering wheel looked much more the part than the standard item. It used the same 'lightning flash' horn push centre as the standard wheel. Dashboards were frequently modified by owners adding additional instruments, switches, map lights or whatever. The flat and generally spacious panel encouraged such attention. Here, in Peter Seamen's Sebring Sprite, an ammeter has displaced the indicator repeater light, the vertical spoke of the wheel most likely obscuring this in its original position in any case.

SPRITE MKI

Notes to table
[1] Iris Blue replaced Speedwell Blue from AN5-9420 (Jan 59).
[2] Leaf Green replaced Dark Green from AN5-9927 (Jan 59).
[3] Beige replaced Primrose from AN5-9605 (Jan 59).
[4] Black seats and trim became an alternative to red from AN5-9603 (Jan 59).
[5] White hood, tonneau cover and side curtains became available from AN5-13543 (Apr 59).
[6] Hardtop available from AN5-8283 (Dec 58).

COLOUR SCHEMES (AN5)

Body colour (paint code)	Seats/ trim	Seat piping	Floor covering	Hood/ tonneau	Hardtop[6]
Speedwell Blue (BU1)[1]	Blue	Light Blue	Blue	Black	Ivory White
Iris Blue (BU12)[1]	Blue	Blue	Blue	Black or White[5]	Ivory White
Dark Green (GN12)[2]	Green	Green	Green	Black	Ivory White
Leaf Green (GN15)[2]	Green	Green	Green	Black or White[5]	Ivory White
Primrose Yellow (YL3)[3]	Black	Yellow	Black	Black	Ivory White
Whitehall Nevada Beige (BG4)[3]	Cherry Red	White	Cherry Red	Black or White[5]	Ivory White
Old English White (WT3)	Cherry Red or Black[4]	White[4]	Cherry Red or Black[4]	Black or White[5]	Ivory White
Cherry Red (RD4)	Cherry Red	White	Cherry Red	Black or White[5]	Ivory White

Both Healey and BMC offered luggage racks for the Sprite. The BMC type is fitted and the Healey version is on the ground. The Le Mans fuel filler cap is another Healey accessory.

This was possibly used on only a few cars and was a forerunner to the BMC sliding pane sidescreens.

Other companies soon followed in producing all manner of equipment for the new Sprite, a basic car which promised huge potential for the accessory industry. It is neither possible nor practicable to discuss everything which was offered worldwide, as this would form a book in itself. But the point I must stress is never to throw away anything which cannot be readily identifiable as being original to a Sprite. In certain instances, a period accessory or modification may just be the most significant and rare part of the whole car. Conversely, many of the Sprites to be seen with disc brakes and wire wheels do not have the original Healey kit fitted. All too often, wheels and brakes have been transferred from a later Sprite or Midget using a Lockheed braking system and larger kingpin/stub axle assembly. It is imperative, therefore, to identify properly any parts which raise a question mark.

COLOUR SCHEMES

BMC's Cowley plant was responsible for paintwork, using ICI synthetic paints. The factory colour schemes are listed in the accompanying table.

Some, if not all, cars had the front frame rails, radiator pillar and the front face of the suspension crossmember flashed over in black. The hood frame for all colour schemes was Cumulus Grey. The underside of the car and the area within the boot space were usually left in brown primer, although some cars were painted body colour underneath. Inside, the floor, tunnel and inner rear wheelarches were sprayed body colour to the highest point only. Underbody protection compound was applied to the rear vertical face of the front wheelarches and to the rear wheelarches. Road wheels were painted Aluminium (Code AL1) with bumper brackets at front (if fitted) and rear in black.

CKD kits assembled in Australia were painted locally in 'international racing circuit colours': Nurburg White, Monza Red, Le Mans Blue and Goodwood Green.

PRODUCTION CHANGES

CHANGES BY CHASSIS NUMBER

501 (Mar 58)
First production Sprite off the Abingdon assembly line on 31 March.

641 (Apr 58)
Copper-gilled radiator core to improve cooling.

1606 (Jun 58)
Side curtain frame retaining screws increased in diameter to /16 UNF. Aluminium door top moulding and frame bracket revised accordingly.

3444 (Jul 58)
Rev counter cable shortened on right-hand drive cars by 1in (25mm).

3689 to 3800 and 4035 onwards (Aug 58)
Speedometer cable shortened on left-hand drive cars by 2in (50mm) to prevent contact with exhaust pipe.

4333, less 4471, 4622, 4680 & 4684 (Sep 58)
Box section rails added to the rear inner wing to axle tunnel and boot floor join. Trim panels and matting revised to suit this modification. Rear shock absorbers repositioned with revised links and rear axle attachment brackets, radius arm brackets enlarged with securing bolts now passing through shock absorbers. Luggage rack becomes an optional extra.

4695 (Sep 58)
Ducting panels added to the bonnet to direct air into the radiator.

4800 (Sep 58)
Improved steering arms fitted, requiring a longer bolt to the forward hole.

4996 (Sep 58)
Clutch pedal and push rod changed.

5321 (Oct 58)
Tooling hole produced in rear crossmember.

5477, and also 5137, 5287 & 5288 (Oct 58)
Windscreen glass, frame and pillars changed, deleting seven hood stud fasteners. Hood revised to suit new attachment method to screen frame. Windscreen wiper motor self-parking cap changed from domed to flat top. Engine compartment inner wings now double-skinned with re-shaped splash shield panels in the bottom of the compartment. Right-hand removable panel now welded to bodyshell. Corner gusset panels added to the top junction with the upper footwell panel. Exhaust pipe and bellhousing bracket revised to suit new splash shield aperture.

6063 (Oct 58)
Rubber buffer blocks added to the tops of the front inner wings to support bonnet.

6136 (Oct 58)
Heater air ducting pipe clip to inner wing panel changed.

6433 and intermittently from 6118 (Oct 58)
Front stub axles revised.

6490 (Oct 58)
Strengthened left-hand radiator pillar.

6652 (Oct 58)
Screw and nut fastenings added to door pocket liner panels (two per panel).

6696 (Oct 58)
Alloy clutch slave cylinder changed to cast iron.

6889 (Nov 58)
Side ducting panels added to radiator and new pressure cap fitted.

6892 (Nov 58)
Horn bracket changed for single horn.

7433 (Nov 58)
Passenger side steering column hole blanking grommet changed.

8283 (Dec 58)
First car fitted with hard top and sliding panes, aluminium-framed sidescreens.

9191 (Jan 59)
Rubber washers replace plain washers behind the bonnet lower front locating cones.

9420 (Jan 59)
Iris Blue replaces Speedwell Blue.

9605 (Jan 59)
Whitehall Nevada Beige replaces Primrose Yellow. Black seats with white piping become an alternative to Red for Old English White cars.

9927 (Jan 59)
Leaf Green replaces Dark Green.

10344 (Jan 59)
Door catch mechanism changed to include shorter handles with chromed knobs.

11126 (Feb 59)
Front suspension wishbone inner fulcrum bolts now secured with Nyloc nuts to replace castellated nuts with split pins.

11760 (Mar 59)
Improved fuel tank filler neck grommet to prevent ingress of water into boot space.

13543 (Apr 59)
White tonneau cover, hood and side curtains introduced as an option according to colour scheme.

15151 (Apr 59)
Road wheel centres now welded instead of riveted to rim. Fuel pipe from tank to pump changed to include a revised nut and nipple at the tank connection.

15553 (Apr 59)
Harder engine mounting rubber blocks and revised float chamber needle valve assemblies to reduce vibration and fuel flooding tendencies.

16151 (Apr 59)
Bonnet safety catch spring changed.

19015 (Jun 59)
Improved sealed beam headlamp units for North American cars.

21546 (Aug 59)
Rubber matting beneath the seats deleted for Cherry Red interiors.

22583 (Jul 59)
Nylon washer added to steering rack pinion shaft to improve effectiveness of felt oil seal.

26139 (Oct 59)
Driver's seat runner revised to increase range of forward adjustment.

26185 (Oct 59)
Rubber matting beneath the seats deleted for blue interiors.

26464 (Nov 59)
Rubber matting beneath the seats deleted for black interiors.

26724 (Nov 59)
Tudor windscreen washer kit changed.

26824 (Nov 59)
Windscreen wiper wheelboxes changed.

Up to engine number 9C-U-H 11888 (February 1959), the ignition coil was mounted on top of the dynamo. It was subsequently relocated to the corner gusset panel.

The Healey sidescreen for use with the company's hardtop is shown in the centre. Top and bottom are prototype sidescreens which Healeys developed in search of an improvement over the original side fabric curtains. The lower sidescreen is a deeper version of that which eventually went into production.

The early windscreen frame stanchion, correctly finished in body colour. The top three screws appear to be missing, while the base screws are modern reproductions with chromed Pozi-drive heads. Cadmium-plated, cross-head drive (Philips) screws would be correct here.

The later windscreen, frame and pillar fitted to all Sprite MkIs from AN5-5477 (October 1958). With this arrangement, the hood was now retained by a sewn-in steel strip locating into a slot in the front of the screen frame. Only the very outer 'lift the dot' fasteners were retained to locate the hood laterally.

27912 (Nov 59)
Improved front suspension wishbones.
31903 (Jan 60)
Fixed passenger seat frame now with riveted brackets to replace screw fittings.
34556 (Mar 60)
Side curtains now listed as being discontinued and all cars now fitted with sliding pane sidescreens as standard. However, a few cars still had side curtains until AN5-39945 (May 60).
35005 (Mar 60)
Radiator cap changed.
36194 (Mar 60)
Number plate lamp unit now contains two bulbs.
37492 (Apr 60)
Rubber matting beneath the seats deleted for green interiors.
39224 (May 60)
Road wheels improved with larger hub ribbing and repositioned vent holes.
41016 (Jun 60)
Coil section of fuel pipe before fuel pump replaced with a flexible pipe to eliminate vibration and chafing.
42200 (Jul 60)
Tool bag changed with revised hub cap lever and spark plug box spanner.
48338 (Oct 60)
Stronger front suspension wishbones with revised spring pans introduced.
48627 (Oct 60)
Steering column top bush changed.
49584 (Nov 60)
Last complete Sprite assembled at Abingdon.
50116 (Feb 61)
Last chassis number allocated as a 'semi-CKD' for an Innocenti Spyder.

CHANGES BY ENGINE NUMBER

101 (Mar 58)
First engine number, but not fitted to first chassis number.
273 (Apr 58)
Oil pick-up suction pipe changed.
320 (Apr 58)
Clutch withdrawal arm rubber gaiter changed.
359 (Apr 58)
Extra fan blade fitted to make a four-blade fan.
1073 (May 58)
Thermostat changed from 75°C to 68°C.
1168 (Jun 58)
New rear main bearing cap to delete drilling and plug.
1169 (Jun 58)
Improved head gasket with ferrules around end water holes.
1397 (Jun 58)
Improved inlet and exhaust valves with modified oil seal and retaining cap.
1551 (Jun 58)
Float chamber lids changed with a copper overflow pipe.
3995 (Aug 58)
Gear lever bent to move knob rearwards by 1⅛in (28mm).

4796 (Sep 58)
Spark plug connector caps now right-angled to replace straight connectors.
4834 (Sep 58)
Flywheel and starter ring gear revised.
5679 (Oct 58)
Changed speedometer drive pinion in gearbox.
6359 (Oct 58)
Modified cylinder block drain tap fitted.
6733 (Nov 58)
Improved cylinder head gasket with ferrules around all the water holes.
10903 (Feb 59)
Solid copper washers, oil filter to cylinder block pipe.
11889 (Feb 59)
Ignition coil removed from the top of the dynamo.
15305 (Apr 59)
Engine mounting rubber blocks changed to a harder specification. Revised float chamber needle valve assemblies.
17365 (May 59)
Carburettor heat shield now with an extension piece to shield fuel pump.
18480 (Jun 59)
Fuel pipe between pump and carburettors changed.
25484 (Oct 59)
Spark plug cap connectors now include a suppressor.
25509 (Oct 59)
Revised anti-rattle plunger for gear lever.
28914 (Dec 59)
Float chamber overflow pipes now with nylon extension pipes.
32663 (Feb 60)
Vacuum advance pipe to distributor revised to delete flame trap.
35641 (Apr 60)
Petroflex pipes between delivery pipe and float chambers changed.
37647 (Apr 60)
Neoprene 'lip' type oil seal replaces felt seal in the timing cover for the crankshaft pulley.
49201
Last engine number allocated.

CHANGES BY GEARBOX NUMBER

946
Changed third and fourth speed synchroniser.
1146
Changed reverse gear selector rod and fork.
11522
Gear lever knob changed, deleting lock nut.
14495
Front selector bush now made in one piece to replace split bush.
18337
Nylon thrust button introduced for rear selector lever.
21267
Speedometer drive spindle bush changed from brass to nylon.

Original Sprite & Midget

SPRITE MkII & MIDGET MkI
(HAN6/7, GAN1/2)

Towards the end of 1960, Sprite MkI sales were slackening. Rightly or wrongly, BMC decided that an update would revive sales and instructed Healey to re-design the front of the car along more conventional lines. Not content with this alone, MG was instructed to re-design the rear, both teams initially working without knowledge of the other!

Fortunately, both parties became aware of this and commonsense saw an overall result which did not look a total mismatch of ideas. Launched in May 1961, the Sprite MkII was a much more user-friendly car, if not quite so visually friendly with its less characterful boxy appearance. Gone was the huge yawning bonnet, replaced by a conventional bonnet separated from fixed wings and a fixed front panel assembly. And gone too was the fear of being swallowed by a giant bug should the bonnet fall shut...

At the rear, a conventional opening boot lid may have spoiled the potholing practice familiar to Sprite MkI owners, but removing the spare wheel or retrieving small objects was no longer a painful task. The new rear bodywork also included an extended cockpit which allowed additional storage space behind the seats. Useful as this was, BMC was a little optimistic if it thought the optional rear seat cushion to fit on the axle tunnel would turn the new car into a four-seater.

Initially, the Sprite MkII was offered in basic or de luxe form, the basic car lacking the de luxe features of front and rear bumpers, rev counter, windscreen washer and adjustment of the passenger seat. In practice, few people ordered the basic version, so BMC soon dropped it and quietly discarded the term de luxe at the same time.

Mechanically, the 948cc engine was improved to produce 46.5bhp, and the gearbox now contained the previously optional close-ratio gear set as standard.

In June 1961, one month after the Sprite MkII's introduction, BMC demonstrated its hand of badge engineering by introducing the slightly more expensive MG Midget MkI. Essentially this was a Sprite MkII de luxe with improved interior trim, better detailed instruments, side waistline and bonnet moulding trim strips, a vertical-slat radiator grille and different badges.

Chassis number prefixes became HAN6 for the Sprite and GAN1 for the Midget, the H now identifying a Healey and G for an MG.

October 1962 saw significant improvements to both

The full rear bumper and overriders of the de luxe specification are fitted to a rare Deep Pink HAN6 Sprite MkII owned by Roy de St Croix since new. Note the body colour fuel cap. A rear bumper was not fitted for the basic specification. Instead, two vertical bumperettes – actually inverted Sprite MkI items – would have been standard wear.

Sprite MkII & Midget MkI

The Sprite MkII (above) looked a totally different car from its predecessor at the front. Only the windscreen gives any clues to parentage. Under the outer wrapper, however, the ingredients were little changed.

Bruce Halliday's Tartan Red Midget MkI (above right) displays the waistline strip and plain hub caps which differentiated the Midget and Sprite in side view. Plain disc wheels without ventilation holes show this to be a 1098cc GAN2 version.

Wire wheels, as seen on Andy Ricketts' GAN2 Midget, were a popular option, although keeping them clean could be an onerous task in winter.

models with the introduction of disc front brakes and a 1098cc engine producing 55bhp. A stronger ribbed gearbox casing contained improved baulk ring synchromesh to replace the cone type. Interior trim was also improved and the cable-driven tachometer was replaced by an electrical impulse type.

Chassis number prefixes advanced to HAN7 and GAN2, but the 'Mk' designations remained unchanged. Only in Australia, where the Sprite was assembled from CKD kits from Abingdon, was the term Sprite MkIIA applied to distinguish the 1098cc version – the Midget was yet to be assembled there.

BODY & CHASSIS

Despite the radical change of appearance, the Sprite MkII and Midget MkI were very much like the Sprite MkI under the skin. The entire floorpan, bulkheads, scuttle, A posts and doors were exactly as before. The changes to the front end were achieved by using bolt-on wings, a new front centre panel assembly, and a new bonnet which retained the existing hinges.

The front centre panel was made up of several pressings welded together to form the bonnet slam panel, radiator ducting and grille aperture, and front valance with stiffener section. Ten ¼ UNF bolts secured the assembly to the front 'chassis' rail extensions and radiator pillars. Nuts were welded into the rails and pillars to suit.

The new front wings were secured by 11 bolts per side, six to the car and five to the front panel assembly. Splash shields, one either side of the radiator, were bolted between the radiator pillars and inner wing support brackets. Owing to the new front wings being fixed, the previously inwardly turned flanged edges of the inner wing were now extended and turned outwards to act as splash guards, in an attempt to prevent road dirt from collecting behind the wings.

The rear end bodywork comprised new wing panels, back panel, shroud panel, inner wing panels and a boot lid. With the exception of the boot lid, all the new panelwork was welded to the main structure.

The new shroud panel allowed for an extended cockpit, the area immediately behind the seats now being open. A Vynide-covered trim panel, set vertically between shroud and rear axle tunnel, closed off the boot space from the cockpit. The extruded aluminium mouldings fitted to the rear cockpit edges were similar to those used in the Sprite MkI, but now of three sections to suit the larger and re-shaped rear cockpit area.

Seat belt mounting points were now included in the rear of the inner sill panels. These took the form of welded bobbins containing the standard 7/16 UNF thread pattern for the seat belt mounting bolts. For the inner mountings, the transmission tunnel was manufactured with holes in raised pressings but without welded nuts. The nuts could be inserted from the propshaft aperture in the rear heel panel from beneath the car. The top mounting was by means of two holes in the top of the rear inner wheelarch, the belt being secured by two nuts from a welded stud plate through the wheelarch.

The spare wheel was now secured to the boot floor by a central bolt engaging into a threaded bracket welded to the floor panel, the Sprite MkI's three strap lugs being deleted. Either side of the floor, extension panels were added to fill the additional width created by the new wings. The boot floor continued to be produced around the original AN5 pattern (tapering to the rear), rather than as a new full-width single panel.

51

BODY TRIM

While the Sprite continued to be simply adorned externally, the new Midget had full-length chromed waistline trim mouldings, and a central chromed moulding on the bonnet.

The Sprite grille, a simple polished aluminium mesh, was replaced on the Midget by a vertical-slatted grille with a chrome-plated Mazak central bar carrying the MG motif at the top. The grille was a complex assembly of individual polished aluminium bars, all screwed to the top and bottom members. Surrounding both types of grille, aluminium finishing pieces were pop-riveted to the bodywork, one beneath the grille and one either side of the grille. The top finisher was riveted to the underside of the bonnet. The finishers were the same for both the Sprite and Midget.

A front bumper bar was standard on the Midget and Sprite de luxe, but an optional extra on the basic Sprite. Without a bumper, two simple steel brackets were employed to mount the front number plate. Front and rear number plates were fitted to black-painted backing plates, varying in size according to market. The rear bumper was also standard on the Midget and Sprite de luxe, the basic Sprite having just two vertically mounted overriders, the same as those fitted to the MkI but inverted.

Front and rear bumpers carried matching overriders with a black rubber mounting strip between the bumper and overriders. All mounting brackets were painted black. The top mounting brackets for the rear bumper were fitted with a chrome-plated capping piece where the top securing bolt passed through the rear panel. The domed nuts were external and chrome-plated.

The bonnet was held open by a black-painted prop which engaged with a slotted plate welded to the left-hand wing drip channel. The prop, pivoted from the front of the bonnet, was stowed in a spring clip towards the rear of the bonnet on the left-hand side. The bonnet rested on four rubber buffer blocks, one in each wing drip channel and two set into the front of the bonnet to rest upon the slam panel.

The boot lid was similarly equipped, with the black-painted prop again hinged to the lid on the left-hand side and stowed in a spring clip towards the rear of the lid. The boot lid was initially secured by an L-shaped handle with a locking barrel, but from chassis numbers HAN6-8692 and GAN1-4582 (October 1961) this was replaced by a T-handle.

The new feature of an opening boot lid must have been accompanied by some development problems, for stronger boot hinges were fitted from HAN6-3548 and GAN1-1209 (June 1961), four rubber support blocks were added to the drip channel from HAN6-8157 and GAN1-4103 (October 1961), and the sealing rubber moulding was changed at HAN6-

The Sprite MkII's bonnet badge sprouted a pair of wings and was to serve unchanged until October 1969.

Lucas lighting with a new side and indicator unit with a separate bulb for each function. Note the shape of the glass lens, which is handed to a particular side of the car. Quite often, a Triumph Herald lens finds its way onto a Sprite or Midget, but this has an apexed rather than curved profile. The new Sprite radiator grille has aluminium surround finishers riveted to the bodywork.

The Midget used a vertical-slat grille of fairly complex construction. The addition of a central bonnet moulding strip further identified the Midget from the Sprite. The front edge of the new bonnet became a notorious rust trap.

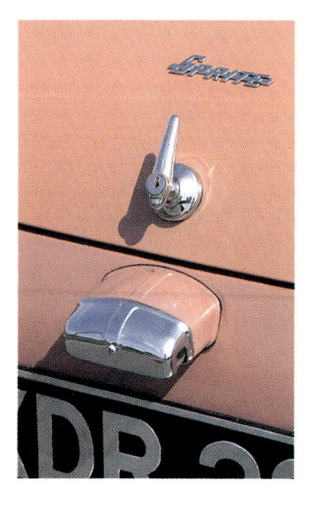

Sprites prior to HAN6-8692 and Midgets prior to GAN1-4582 (October 1961) used this L-shaped boot handle (left). The number plate light unit was the same as that used on the MkI Sprite but mounted on a new Mazak plinth. The Midget upstaged the Sprite with much more elaborate boot lid badging (right), while the later T-shaped handle is shown here.

SPRITE MKII & MIDGET MKI

The hood frame was similar to that of the 'Frogeye', but could now be split at the centre for storage in the boot, a bag being provided.

A safety catch prevented the bonnet from lifting should the centre latch not engage. Note the rubber buffer seating and aluminium finisher riveted to the lower edge.

The new Lucas rear lamp unit. The fuel cap was still body colour unless the optional Wilmot Breeden chromed locking cap was ordered.

The hood followed the same pattern and means of fitment as the 'Frogeye'. The sidescreens now allowed both perspex panes to be moved.

17637 and GAN1-11890 (March 1962). Indeed, BMC did not recommend fitting the optional luggage rack to an earlier model unless it was fitted with the stronger hinges and rubber supporting blocks.

All was not well at the front either, the Midget's central bonnet moulding requiring an extra securing stud from GAN1-1508 (July 1961) to ensure that it remained in close contact with the bonnet profile.

The Sprite continued to wear a bonnet badge, similar to that of the MkI but now sporting a pair of wings. The badge was now secured to the bonnet by four tabs located through slots in the bonnet pressing and retained by spring clips on the inside. The

Midget carried no bonnet badge, front end identity being given by the MG motif in the top of the radiator grille. On the boot lid, a simple 'Sprite' script was carried on a single-piece chromed badge placed centrally above the handle. The Midget used rather more complex badging, with a chromed octagon surrounding separate MG letters and, below this, a single-piece 'Midget' script badge. A total of 15 holes and spire clips were required to secure these four separate pieces to the boot lid.

WEATHER EQUIPMENT

The hood and optional tonneau cover were available in a new range of colours, while the hood frame, painted Cumulus Grey and similar to the Sprite MkI design, could now be split in the centre of the two bows. The frame could only be stowed in the boot as no stowage location boxes were now provided on the heel panel behind the seats; a bag was provided for the frame halves.

Because of the longer cockpit, a Cumulus Grey tonneau support rail, again split at the centre, located into the hood frame support tubes behind the tops of the B post; this shared the hood stowage bag when not in use. The tonneau cover now had sewn-in flaps running across the cover, either side of the central zip. Each flap contained two 'lift the dot' fasteners which engaged with pegs on the heel panel to anchor the tonneau behind the seats while still covering the rear cockpit area.

The sidescreens now contained two sliding perspex panes, each pane having a small glued-on block of perspex to act as a handle. As before, a stowage bag contained both sidescreens when not in use.

The optional hardtop for the Sprite was initially a revised version of that offered for the MkI. The Midget, however, used a completely new design with a flat rear window and a more angular appearance. This option required a different sidescreen with a rearwards-sloping rear edge, whose angle aligned with the door's rear edge. The Sprite hardtop was now in Old English White, whereas the Midget version could be ordered in this colour as well as red, blue or grey. Both hardtops were superseded by a common design based upon the Midget version but using the standard sidescreens. The owner of an earlier Midget, therefore, would have required two sets of sidescreens to suit both hood and hardtop.

INTERIOR TRIM

Upon its introduction, the Sprite MkII was available in basic or de luxe specification, the de luxe having an adjustable passenger seat.

The seats were carried over from the Sprite MkI, but with the addition of two new colour options – red with black piping and black with red piping. The

Stowage bags are, top to bottom, Vynide for hood frame and tonneau sticks, Vynide for sidescreens, grey cotton twill for tonneau cover and grey cotton twill for hood. Not shown is the Vynide tool kit bag.

The tonneau cover could be fastened in this position by folding it over the support rail and engaging the 'lift the dot' fasteners with studs screwed to the heelboard. The centre zip allowed the passenger side to be fully covered while the driver braved the elements with the cover stowed behind his or her seat.

All neatly stowed away, the rolled hood in its storage bag secured by two straps to the divider panel. The hood frame tonneau sticks are split and stored in the black bag around the wheel. The wheel is secured to the floor by a central bolt. The fuel filler neck would have been finished in cadmium plating.

The ribbed rubber door pocket lining of the 'Frogeye' was replaced by Vynide for the Sprite MkII and Midget MkI.

The seats of the HAN6 Sprite MkII were basically unchanged from those of the MkI, although the passenger seat could now be adjusted for reach on the de luxe version. Note the flecked carpeting and optional rear seat cushion.

Sprite MkII & Midget MkI

The interior of Joy Dowling's remarkably original GAN1 Midget shows the 'upmarketing' of the seat coverings over the humbler HAN6 Sprite.

The tonneau rail fitted into the hood frame sockets, and was divided in the centre for stowage in the boot. A bag was provided to keep both halves safe when not in use. The optional rear seat cushion matched the front seat pattern and colour.

seat coverings were revised at the following points: blue with blue piping, HAN6-6285 (September 1961, changed to the same shade of blue as used in the Midget GAN1); black with red piping, HAN6-6723 (September 1961); red with white or black piping, HAN6-14842 to 14965, 14980 to 15211 and 15219 onwards (February 1962). Green seats with green piping appeared on just 13 GAN1 Midgets with Almond Green paintwork. In early 1962 black with white piping continued over from the Sprite MkI until the introduction of the Sprite MkII (HAN7) in October 1962.

The Midget MkI was provided with different seat coverings upon the same framework and runners as

As with the HAN6 Sprite MkII, the GAN1 Midget MkI – seen here – had flecked carpet over the axle tunnel and rubber matting on the floor, but flecked carpet on the lower portions of the seat backrests was special to the Midget.

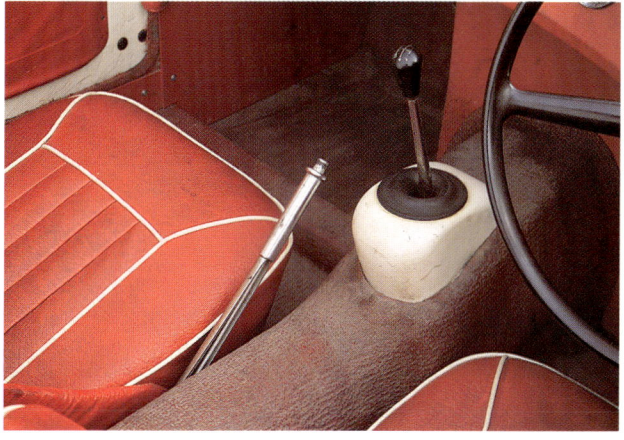

This very original Midget's matting has faded over the years, but is complete and undamaged. Note the rubber's textured surface.

55

Andy Ricketts' Midget GAN2 with Hazelnut trim. The dashboard has a number of non-original additions, but it does serve to show the presence of a padded roll to the lower edge. This change also applied to the Sprite HAN7, which gained the upper padded roll at the same time. The gear lever gaiter is incorrect and the metal surround should be Old English White to match the exterior.

used in the Sprite, both seats being adjustable. The fluted area of material was reduced with a plain border surrounding the sides and front of the cushion, and the sides and top of the squab. The lower rear portion of the squab was now covered in flecked carpeting to suit, or to contrast with, the main trim colour. Details of colour-coding and revisions are given in the 'Colour Schemes' table at the end of this chapter.

With the introduction of the HAN7 and GAN2 models, the seats for both Sprite and Midget were commonised to a new pattern in a revised colour range of red, black, blue or hazelnut. Both cushion and squab were now given a single transverse pleat across the fluted section to add a little extra shaping, but with no apparent improvement in comfort. Flecked carpeting was discontinued to leave the carpeted section on the rear of the seat squab in the main trim colour. The transverse pleat on the cushion disappeared when a new cushion was introduced at the following points: hazelnut, HAN7-31701 and GAN2-20941 (May 1963); black, HAN7-32076 and GAN2-21041 (May 1963); red, HAN7-32095 and GAN2-21091 (May 1963); blue, HAN7-32148 and GAN2-20886 (June 1963).

An option of a rear seat cushion became available with the lengthened cockpit of the HAN6/7 and GAN1/2. Placed on the axle tunnel, it was held along the rear edge by four self-tapping screws and at the front by two straps containing 'lift the dot' fasteners to engage with pegs screwed to the rear panel behind the seats. Available in corresponding trim colours, the cushion was again a dedicated pattern on HAN6 and GAN1 versions but a common design for HAN7 and GAN2.

Interior trim liner panels for HAN6 and GAN1 were based upon those fitted to the Sprite MkI. However, due to the reshaping of the rear cockpit, an additional vertical divider panel separated the rear of the cockpit from the boot. The rear edge of the cockpit was again finished with extruded aluminium mouldings, now comprising a Vynide-covered centre piece and two untrimmed side pieces. For HAN7 and GAN2, all three mouldings were covered in Vynide, the centre one being padded.

Similarly, the scuttle top moulding was covered in Vynide on the HAN6, but on the Midget GAN1 this became a padded roll – a feature which the Sprite did not receive until the start of HAN7 production. Both Sprite HAN7 and Midget GAN2 models received a corresponding padded roll to the lower edge of the dashboard panel.

The door skin liner, of ribbed rubber on the Sprite MkI, was replaced on all Sprite MkIIs and Midget MkIs by a Vynide covering. For HAN7 and GAN2 versions, the painted inner door frame also received a Vynide-covered liner panel and the door top moulding was covered in Vynide to match the interior trim.

As with the GAN1 Midget, carpeting continued to be applied to the lower portion of the seat backs on the GAN2 version, possibly to prevent damage to the Vynide when the optional rear seat cushion was occupied.

Both Midget GAN2 (right) and Sprite HAN7 gained door liner panels to hide painted metal. This door pocket appears a little overpadded. David Oliver's HAN6 Sprite (far right) shows the close connection with the 'Frogeye'. However, the toggle switch, repositioned heater control, fuel gauge markings, ignition and main beam warning lights are all revisions.

Prior to GAN1-5034 (November 1961), the Midget MkI featured a creamy white steering wheel and cowl, but it is incredibly rare for these features to have survived. Note the different style spokes when compared to the equivalent HAN6 Sprite MkII steering wheel. Jaeger instruments were better detailed than those of the Sprite. Carpeting is not correct, for flecked rubber matting would have been original to this model. Healeys offered a carpet set, but this partially overlapped the gear lever cowl.

Floor coverings for HAN6 and GAN1 were rubber matting, that in the Midget containing a fleck appropriate to the colour scheme. Carpeting was applied to the axle tunnel, both Sprite and Midget having a flecked pile. All floor and axle tunnel coverings on HAN7 and GAN2 versions were commonised with full plain-coloured carpeting, the front carpets containing inset rubber mats for driver and passenger heel areas.

The boot space had no floor covering or side trim panels. The spare wheel was stowed outer hub downwards in the case of a steel disc wheel, or outer hub upwards for a wire wheel. The wheel was secured by a central bolt and large washer, a welded nut being provided on the boot floor.

DASHBOARD & INSTRUMENTS

The dashboard, instruments and controls for the HAN6 Sprite MkII were slightly revised from those of the MkI.

The ignition switch, still placed centrally in the panel, no longer included a surrounding bezel control for the lights, a new three-position toggle switch immediately to the right of the ignition switch now providing this function. A matching two-position toggle switch to the left of the ignition switch controlled the single-speed, self-parking wipers. Neither switch carried any marking to identify its function.

Placed immediately above the ignition switch, the flashing indicator switch was the same as before, with the windscreen washer control to the left and the choke control beyond this again. To the right of the indicator switch was the repositioned heater control, with the starter pull switch to the right of this again. Choke, washer, heater and starter controls were appropriately marked. On left-hand drive versions, the heater and windscreen washer controls were transposed with the choke and starter controls, putting the starter within easier reach of the driver. For left-hand drive cars with a combined steering lock and ignition/starter switch, the dashboard-mounted ignition switch and starter pull control were deleted. The green indicator warning light, instrument lighting and speedometer trip meter reset control were all carried over from the Sprite MkI.

The revised instruments, placed similarly to those

This view gives a few more clues to distinguish between Sprites AN5 and HAN6, for the latter has toggle switches for lights and wipers. The ignition and main beam warning lights swap positions and now have jeweled lenses, in red for ignition (as before) but now blue for main beam.

Original Sprite & Midget

The Midget GAN1 dashboard and steering wheel after GAN1-5034 (November 1961). The wheel and surround were now finished in black, but the wheel retained the thinner spokes unique to the Midget.

in the Sprite MkI in both right- and left-hand drive forms, carried the following Smiths identification numbers: combined oil pressure (upper) and water temperature (lower) gauge, GD1502/06 (Fahrenheit) or GD 1502/08 (Centigrade); rev counter (with red ignition warning light), RN2313/00; speedometer (with blue main beam warning light), SN6126/00 (mph) or SN6126/01 (kph); fuel gauge, FG2530/64. Compared with the Sprite MkI, the ignition and main beam warning lamps swapped positions within the rev counter and speedometer, and used larger 'jeweled' lenses.

The passenger side of the facia panel still contained a chromed grab handle with the optional radio fitted below this in a push-out section of the panel. Under the left-hand side of the dashboard, the bonnet release, marked 'B', was carried in a small bracket screwed to the side of the inner footwell panel.

The GAN1 Midget MkI differed from the Sprite in having Jaeger instruments with more detailed markings. Despite the name, these instruments were still made by Smiths and carried the following identification numbers: combined oil pressure and temperature gauge, GD1501/10 (Fahrenheit) or GD1501/11 (Centigrade); rev counter, RN2312/01; speedometer, SN6125/02 (mph) or SN6125/03 (kph); fuel gauge, FG2530/63. The Midget passenger was not provided with a dashboard-mounted grab handle. Instead, an octagonal MG motif with horizontal lines from either side was applied to the panel. One further feature separated the Midget from the Sprite. Up until chassis number GAN1-5034, the Midget used a cream steering wheel with a dashboard-fitted steering column surround in matching cream plastic. After this point, the steering wheel and steering column shroud became black.

With the introduction of the 1098cc engine for the Sprite MkII (HAN7) and Midget MkI (GAN2), revisions were made to the instruments and dashboard panel. The cable-driven rev counter was replaced by an electrical impulse type. The instruments on both cars assumed a commonised Smiths pattern as follows: combined oil pressure and temperature gauge, GD1501/12 (Fahrenheit) or GD1501/13 (Centigrade); rev counter, RV12401/00; speedometer, SN6135/00 (mph) or SN6135/01 (kph); fuel gauge, FG2530/70. A padded roll in interior trim colour was added to the lower edge of the dashboard.

Engine

Although remaining at 948cc, the A-series engine for the Sprite MkII (HAN6) and Midget MkI (GAN1) received a number of improvements. Power output rose to 46.5bhp at 5500rpm, but with only a slight increase in torque to 52.8lb/ft at 3000rpm. With more weight to pull along and a natural desire to

Excuse the condition of the bezels, but the mileage is true. Joy Dowling has owned her Midget since new and always drives it with the roof off.

58

Sprite MkII & Midget MkI

SU 1¼in HS2 carburettors and paper element air filter casings are the most noticeable changes for the 9CG 948cc engine of the HAN6 and GAN1 models. Internal improvements boosted power to 46.5bhp at 5500rpm. The filter casings are without the Cooper script and numbers in this view.

An internal bonnet release knob was provided for both Sprite and Midget.

increase performance of the new version, the engine now became much more specialised when compared to the lightly-modified Austin A35 unit used in the Sprite MkI.

This higher-specification engine carried the prefix 9CG, which was now followed by a new prefix for the transmission. The U disappeared, to be replaced by DA, which signified that the gearbox now had a close-ratio gear set. The final letter of the prefix was either H for high (9:1) compression or, for the first time, an optional L for low (8.3:1) compression. As with the 9C engine, the 9CG sequence began at 101 and continued to 36711. Gold Seal replacement engines of the 9CG type had prefixes of 8G16 (high compression) or 8G17 (low compression).

Sprites and Midgets drew their engines from the same numbering sequence, so no two numbers were the same for either application. The only identifying difference was the rocker box plate: the Sprite continued with an Austin plate riveted to the right-hand side while the Midget had a similar plate reading MG. Despite the more traditional practice of MGs having red engines, the Midget shared the corporate 'engine green' used on all A-series engines.

The high compression ratio must have caused some concern to BMC, as the company saw fit to print a warning in the owner's handbook to the effect that the engine may be extremely sensitive to variations in fuel, ignition timing and the heat range of the sparking plugs. To worry the owner even further, the handbook also stated that failures were bound to occur if the special recommendations concerning these items were not adhered to, and that regular decarbonising of the cylinder head should also be carried out if damage was not to occur. Happily, BMC's words of warning did not appear to matter too much, for the engine proved to be both eager and reliable in service despite the antics of many an owner.

The higher compression ratio was achieved by using flat-topped pistons, whereas the lower 8.3:1 ratio used dished-crown pistons similar to those in the 9C engine. Oversize pistons of +.010in (.254mm), .020in (.508mm), .030in (.762mm) and .040in (1.016mm) were available in both flat and dished types; cylinder liners could be fitted when further overbore sizes were not possible.

A new camshaft providing a valve lift of .312in (7.925mm) now ran in three replaceable thin-wall copper-lead bearings. It was timed to give an inlet opening at 5°BTDC and closing at 45°ABDC, and an exhaust opening at 51°BBDC and closing at 21°ATDC. The cylinder head contained larger inlet valves of 1.15in (29.2mm) diameter, but the exhaust valves remained at 1.00in (25.4mm) diameter. Double valve springs were now fitted to each valve to allow the use of higher revs.

Despite BMC's concerns about engine durability, no significant modifications were required during production, although the oil filter arrangement was altered from engine number 9CG-DAH-5067. This now reverted to the A35's vertical mounting type, but a new adaptor between cylinder block and filter head raised the filter sufficiently to allow the element to be changed without interference from the main frame rail. Even so, this new position for the filter made element changes more difficult than with the previous angled filter canister. The oil feed pipe between filter head and cylinder block also had to be changed.

A steel crankshaft timing gear sprocket replaced the cast iron gear from 9CG-DAH-5093, and an improved rocker shaft was fitted from 9CG-DAH-12256 and 9CG-DAL-10996. The previous alternative Burman vane type oil pump was discontinued at 9CG-DAH-15258 and 9CG DAL-14245 to leave just the Hobourn-Eaton concentric type pump, but the alternatives of Tecalemit or Purolator oil filter assemblies remained.

From October 1962, the first capacity increase took effect from chassis numbers HAN7-24732 and GAN2-16184. The bore and stroke were increased to 2.543in (64.58mm) and 3.296in (83.72mm) to give a capacity of 1098cc. The new prefix was 10CG-DA, and the compression ratio was 8.9:1 (H) or 8.1:1 (L). Power output in high compression specification was 55bhp at 5500rpm with torque of 62lb ft at 3250rpm. The numbering sequence was 101 to 21048, shared by both Sprite and Midget.

The longer-throw crankshaft retained the 948cc engine's main and big end bearing journal sizes, which was to prove a limiting feature of the uprated engine. The crank suffered from a rigidity problem which caused vibration periods at certain engine speeds and loadings. BMC were certainly aware of the shortcomings, and Special Tuning did not offer any performance-increasing equipment as a result.

Gudgeon pins were now a fully floating type with the deletion of connecting rods with clamp-bolted small ends. The connecting rods now had replaceable press-fitted bushes. The solid skirt pistons were fitted with circlips to retain the gudgeon pins.

From engine number 10CG-18628 (given as 14878 in the factory parts book, which may be incorrect), the inlet valves were increased in size from 1.153in (29.36mm) to 1.22in (31mm), along with re-shaped combustion chambers. Any increase in power was not stated for this improvement. The Hobourn-Eaton concentric type oil pump was fitted but with alternative part numbers of 12G 50 or 12G 793.

To improve the attachment of the inlet manifold to the cylinder head, spigot bushes were fitted into counter-bores produced in both the head and manifold in both inlet ports. This certainly took effect at 10CG-18628, the relevant BMC service bulletin to dealers claiming that the inlet valves and combustion chamber revisions also took place at this time, in variance with factory parts book information. This modification ensured better matching of the ports and relieved the mounting flanges of stress which could result in cracks forming.

The rocker cover was now fitted with a black plastic oil filler cap with a retaining strap around the oil filler neck. The crankcase ventilation pipe fitted to the front tappet chest cover was reshaped with an upward turn in a forwards direction, this was to act as a means of reducing the amount of oil likely to be directed into the lower vent pipe. This was now secured with a 'P' clip to the front fuel pump stud instead of the rear stud.

Overbore sizes for this engine were only listed as +.010in (.254mm) and .020in (.508mm). The higher compression pistons now reverted to a dished top, with the low compression alternative having a deeper dish. A cylinder liner was still listed as a service item to salvage an excessively worn or oversize cylinder bore. Gold Seal replacement engines carried the part numbers 8G 124 (high compression) and 8G 125 (low compression), or, with the larger inlet valves from 10CG-18628 onwards, 8G 135 (high compression) and 8G 136 (low compression).

COOLING SYSTEM

The cooling system was unchanged save for a revised thermostat water outlet housing for the 1098cc engine. The thermostat was either a bellows or downward-opening wax type, the latter being changed to an upwards-opening type from 10CG-15607. The opening temperature varied according to the operating conditions, as follows: 73°C (163°F), hot climate (948cc engine only); 74°C (165°F), hot climate; 82°C (180°F), temperate climate; 88°C (190°F), cold climate.

CARBURETTORS & FUEL SYSTEM

A new cast alloy inlet manifold was introduced to suit the twin 1¼in HS2 SU carburettors which now replaced the H1 type.

The general layout was similar to before, but with a number of detail revisions. The choke cable now operated on the rear carburettor, with a rod and lever transferring movement to the front carburettor. The jet tubes received fuel via a flexible pipe from the base of each float chamber.

The AC Sphynx mechanical fuel pump delivered fuel to the front float chamber through a Smiths Petroflex pipe. The float chamber lid was provided with a second outlet stub to feed the rear carburettor float chamber through another Petroflex pipe. This pipe was secured by 'P' clips to the upper carburettor mounting studs. Some early versions may have continued to use a rigid steel fuel pipe similar to that used on the Sprite MkI. No float chamber overflow pipes were fitted; instead, a small vent hole beneath the inlet stub was shielded by a small aluminium baffle

Voltage regulator and fusebox were re-located for the Sprite MkII and Midget MkI. A wire clip should secure the air ducting pipe to the blower casing.

plate. A small identification plate was attached to each float chamber lid, secured by one of the three retaining screws. The front carburettor body had a stub pipe connection for the vacuum unit.

These HS2 carburettors did not contain a throttle return spring, so an external spring was connected to each carburettor throttle lever, the lower end of the spring being anchored to the bottom of the revised heat shield. A further spring placed between these acted directly onto the throttle cable lever. The outer cable was located in a revised abutment at the top of the heat shield.

Dashpot damper tops were of black plastic with a scalloped edge, replacing the hexagonal brass tops of the H1 carburettor. Fuel metering needles were types V3 (normal), V2 (rich) and GX (weak). The dashpot piston control spring was colour-coded blue.

Two pressed steel Cooper air filter casings contained renewable paper element filters to serve each carburettor. The filters were secured to each carburettor body by two 4½in long 5⁄16 UNC bolts passing through each casing, and a support tube was fitted to each bolt inside the casing to prevent it from collapsing as the bolts were tightened. A 5¾in long intake nozzle on each filter collected air, the front one directed forwards, the rear one rearwards and slightly upwards. The front filter was fitted with a stub intake for the breather pipe from the rocker cover. A small drain hole was drilled into the lowest point of each casing. Care had to be taken not to refit the filter backplate upside down when replacing the paper element, as the dashpot vent hole in the front face of the carburettor could otherwise be blocked off.

The filter casings were connected together with a tab from each, a stud from one tab passing through a corresponding hole in the other tab. A lock nut then secured the two casings together. The filter casings were painted black with yellow lettering which referred the owner to the vehicle manufacturer's handbook for the servicing of the paper elements. The wording was printed on the top of each casing to read from the left-hand side.

Fuel tank capacity remained as before, but the filler neck was now reduced in length to finish inside the boot space. A flexible and angled connector was mated to a new, additional filler neck leading to the repositioned fuel cap in the rear body panel on the right-hand side of the number plate.

With the introduction of the 1098cc (10GC type) engine in the HAN7 and GAN2 models, the carburettor fuel metering needles were changed to types GY (normal), M (rich), and GG (weak). The fuel system otherwise remained unaltered until engine number 10CG-DA-18628 (H or L), when the inlet manifold was revised to accept spigot locations in each inlet port. Both the cylinder head and manifold were counter-bored concentrically to the inlet tract to allow for the fitting of a split location spigot. This not only provided positive alignment of the inlet manifold to the cylinder head but also relieved the flange faces of the manifold from much of the loading and vibration that could lead to a flange cracking.

Undeservedly, twin carburettors earned a reputation for being difficult to keep in tune and requiring frequent attention. In reality, this was more often due to an owner or mechanic not properly diagnosing another more basic fault, and twiddling with the carburettors in the vain hope that a miracle might happen. The twin SU carburettors are very reliable and do not go out of tune overnight, only accumulated wear over a long period normally causing the settings to become noticeably upset. Providing the dashpot oil levels are kept correct and the air filter elements are not allowed to become choked, there are only three main causes of poor carburation.

First, very high mileage can cause wear to occur on the brass throttle spindles where they pass through the main body. This will allow additional air to be drawn in at small throttle openings, creating a weak mixture. Richening the mixture at idle speed will correct this, but cause over-enrichment at wider throttle openings. The mixture strength, therefore, should not be adjusted until the condition of the throttle spindles is checked. In practice, the spindle wears far more rapidly than the carburettor body, so a replacement spindle often provides a complete cure.

Second, the tapered fuel metering needle can also become worn, especially if the jet assembly has not been correctly centred to the needle. Should the needle and jet rub, wear soon takes place and causes an over-rich mixture at low throttle openings, when the thicker part of the needle rubs against the jet. Again, compensating for a rich mixture at idle speed will cause problems at wider throttle openings.

Finally, the float chamber needle valve assembly can wear, allowing the fuel height in the chamber to become excessive. This again causes an over-rich mixture and, in extreme cases, flooding and an external fuel leak through the vent hole.

BMC became aware of the problem of excessive vibration being transmitted to the float chambers, causing fuel frothing and flooding as the needle valve failed to seat properly. A service memorandum advised dealers to change the rubber mounting grommet between the carburettor body and float chamber to an improved type with better vibration absorption. A change of metering needle from a GY to AN was also recommended.

The need for frequent adjustment of carburettors indicates that something is worn and should be replaced. Once set up correctly and in good working order, the carburettors will maintain their state of tune for many thousands of miles. By the same token, the mechanical linkage between the carburettors should not need adjustment once set correctly.

EXHAUST SYSTEM

A revised exhaust manifold casting for the 9CG engine removed the 'hot spot' connection face. This appeared a more efficient shape than before and may have been partly responsible for the increase in power of the 9CG engine. This manifold also served the 10CG engine.

The exhaust pipe and silencer were unchanged from the Sprite MkI, although a modified system was introduced to comply with European regulations.

The factory parts book illustrates a short extension pipe fitted to the silencer to prevent exhaust gases discolouring the chrome bumper, but this was possibly deleted at a later stage by a longer tailpipe being employed. Initially, the rear bumper bar for the HAN6 Sprite MkII was listed as an optional extra: without the bumper, the tailpipe would have protruded rather too far from the rear of the car and caused a hazard, so one can assume the extension piece would not have been fitted for this application.

ELECTRICAL EQUIPMENT & LAMPS

The wiring loom's black cotton covering now had a white or yellow fleck. The battery spill tray was changed at HAN6-7960 and GAN1-3734 (October 1961) for a new item in cream plastic with six recesses around the sides. The battery retaining bar and rods were changed at HAN7-27756 and GAN2-18220 (January 1963) to suit the change to a Lucas N9 or NZ9 (dry-charged for export) battery.

The voltage control box, type RB106/2, was relocated to the bulkhead panel to the right of the battery. The fuse box, now a Lucas 4FJ, was also relocated to the top footwell panel just below the control box. Both of these were now provided with spade connections, as were the coil LT tappings, flasher unit, dynamo, brake light switch, dip switch, windscreen wiper motor and fuel tank sender unit. Both dynamo and control box were revised from HAN6-7499 and GAN1-3491 intermittently, and permanently from HAN6-10118 and GAN1-7520 (December 1961). The starter motor bendix gear was changed from a nut to a circlip retention at an unspecified time on the 9CG engine.

With the introduction of the 1098cc (10CG) engine, the rev counter was changed from a cable drive to an electrical impulse type. The dynamo was now changed to a C40-1 type without provision for the cable-drive gearbox on the backplate. The dynamo was changed again at 10CG-DA-2571, the later type recognisable by having a plain front end plate instead of a ribbed one.

The windscreen wiper motor, type DR3A, was changed at HAN6-6276 and GAN1-3007 (RHD, August 1961) and at HAN6L-5660 and GAN1L-2206 (LHD, September 1961).

The distributor on the 948cc engine became a DM2 P4 unit, while the 1098cc engine used a type 25D4. A revised distributor for the 9CG engine was introduced at H27445 and L16949, while a further revision occurred at an unspecified point for the 10CG engine. Both capacities used Champion N5 spark plugs with a Lucas LA12 ignition coil. HT leads for the 9CG engine were lengthened for number 1, 2 and 3 cylinders and shortened slightly for number 4 owing to the distributor being rotated anti-clockwise in its housing from H8985 (plus 8690-8700) and L1043. This also required a shorter vacuum advance pipe to be fitted to suit the more upright position of the connection on the vacuum diaphragm unit.

The 10CG engine received new suppressor caps for the spark plugs, while for the French market the HT leads and caps were of an improved type to provide suppression to local regulations. This also called for the distributor to be fitted with a steel suppression screen, the top of the screen being released by two wing nuts to provide access to the distributor cap for servicing.

For some European markets, an electric starter solenoid replaced the cable-operated type, but would possibly not be found on cars prior to July 1961. The solenoid was key-operated from the combined steering lock, ignition and starter switch mounted on the right of the steering column.

Headlamp specifications varied according to local regulations, and were changed during production. The change points of headlamp assemblies and chrome rims, along with an accompanying change in seating rubber gasket, are given in the 'Production Changes' list later on. Headlamp bulb ratings were as follows for main beam and dipped (vertical dip): LHD except North America and Europe, 42W/36W; LHD Europe except France and Sweden, 42W/40W; LHD France and Sweden only, 45W/40W. Sealed beam headlamp units were fitted for home, RHD export and North American markets as follows: home and RHD export, 60W/45W; North America, 50W/40W.

Front side light and indicator units now contained a separate lamp for each function in a new rectangular housing; the indicator lamp was shrouded by an amber lens. For countries where a white flashing indicator was specified, the unit was fitted with a single twin-filament bulb and a glass lens with just one set of concentric diffuser rings. The lamp units were left- and right-handed only when separate bulbs were used for side and flash, the lens having two sets of concentric diffuser rings.

New rear lamp assemblies contained a lower twin-filament bulb for side and brake lights and, above this, a single-filament bulb for the flashing indicators. The light lens, type L676, was formed in two pieces which slotted together and were retained to the backplate by a chromed rim piece. The upper lens for the indicator was either amber or red to suit local lighting regulations. The lower red lens also contained the reflector 'reflex' at the bottom in a raised section. The lenses were very similar to those used on the BMC 1100/1300 model but are not compatible, a point worth remembering at autojumbles as some owners have found to their cost. The number plate light was the same twin-bulb unit as before, but it was mounted on a new Mazak plinth painted body colour.

TRANSMISSION

The 9C-U-H clutch was retained until engine number 9CG-DA-2140, when an uprated pressure plate and cover assembly were fitted. Clutch diameter remained at 6¼in (160mm), but with an increase of total spring pressure to 630/690lb (286/313kg). With the introduction of the 1098cc engine (10CG-DA), clutch diameter was increased to 7¼in (184mm).

The 948cc gearbox now adopted as standard the previously optional close-ratio gear set giving the following ratios: first, 3.2:1; second, 1.916:1; third, 1.357:1; fourth, 1.0:1; reverse, 4.114:1. The input shaft also received a needle roller bearing to support the nose of the mainshaft, to replace the plain phosphor bronze bearing used in the Sprite MkI gearbox. The gearbox front cover and clutch withdrawal arm were changed at 9CG-DA-2140 in line with the uprated clutch being introduced. The synchromesh cones were improved at gearbox number 20330.

With the 1098cc engine came a new raised-rib pattern gearbox casing to improve strength and reduce noise. The top of the bellhousing was now produced with an aperture, fitted with a rubber dust-excluder. The forward gear ratios remained the same but reverse gear now became 4.12:1. Baulk ring synchromesh now replaced the slightly ineffective cone type, but was still omitted from first gear.

The propeller shaft remained the same as that used in the Sprite MkI.

REAR AXLE

No changes to the axle assembly occurred until the introduction of the 1098cc engine for the Sprite HAN7 and Midget GAN2 versions in October 1962. The differential carrier was revised with the deletion of the oil filler/level plug on the right-hand side. In practice, this plug was never required as the axle casing had always been provided with a similar plug in the rear of the casing.

With the introduction of the larger engine, wire wheels became an optional alternative to steel disc wheels, taking effect from body numbers ABL-20812 (Sprite) and GBE-16780 (Midget) – I stress that *body* number rather than *chassis* number is used to define this point. Fitting wire wheels necessitated using a narrower axle casing along with correspondingly shorter handbrake actuating rods. The half shafts were now produced with splines at both ends, the outer end being a press fit into the splined hub adaptor. The hub bearings and brake drums were common to steel and wire wheel axles, although the hub studs were shorter on the wire wheel axle as these were required to secure only the hub and brake drums, not the road wheel.

The axle ratio remained at 4.22:1 for both 948cc and 1098cc engines, although BMC Special Tuning continued to offer alternative ratios.

FRONT SUSPENSION

No changes to the front suspension were made for the HAN6 or GAN1 models other than to fit a larger diameter bump stop rubber buffer within the coil spring from HAN6-20820 and GAN1-13763 (May 1962).

With the introduction of disc front brakes in October 1962 for the Sprite (HAN7) and Midget (GAN2), revised kingpins and new forgings for the stub axles were required. The kingpin lower bearing diameter was increased from .687in (17.45mm) to .780in (19.81mm), but the top bearing remained at .624in (15.85mm) diameter. The new stub axle was drilled with two 7⁄16 UNF holes at the trailing edge to mount the Lockheed brake caliper, the leading edge having a single 5⁄16 UNF threaded hole to secure the brake dust shield. The stub axles now became left- and right-handed components, whereas the previous drum-braked stub axles were interchangeable.

An anti-roll bar became an optional extra from January 1964. This was of 9⁄16in (14.3mm) diameter and differed in detail from the Healey Motor Co Ltd anti-roll bar kit which first appeared on the Sprite MkI. The bar was anchored beneath the front frame rails, in rubber blocks. These were held in place by a steel strap and two bolts engaging into a rectangular threaded mounting block welded beneath the front of each frame rail. As these mounting blocks were

Ventilated wheels were replaced by plain steel disc wheels from HAN7-26240 and GAN2-17166 (December 1962). These proved a stronger design and were recommended by BMC Special Tuning to be fitted to earlier cars used in competition – size was still 3½-13.

Squared-off rear wheelarches with a small flare still contained ventilated steel disc wheels on the HAN6 and GAN1 models, and early HAN7 and GAN2 models. The Sprite retained the 'AH' hubcaps whereas the Midget had plain ones.

BMC offered the Ace Mercury wheel disc by Cornercroft for many of its cars. Whether this seemed appropriate for the 'Spridgets', only the customer could decide. Three types were available.

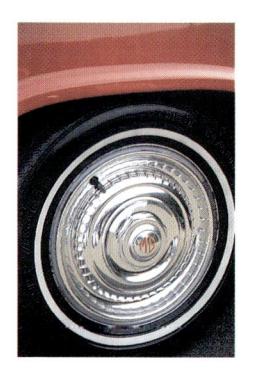

included from the start of Sprite MkI production, the provision of an anti-roll bar had obviously been considered by BMC but not implemented until now.

To keep the anti-roll bar located transversely, two split-collar clamps were provided on the bar to the inside face of each rubber block. The outer ends of the bar were flattened and drilled to take a short drop link with a rubber-bushed stud in each end. The upper end of each link was secured to a triangular plate which was in turn held to the front of each wishbone by three bolts and nuts.

All wishbones were now pre-drilled to accept the optional anti-roll bar attachment plates. Both front and rear faces of the wishbone were drilled to retain the interchangeability of wishbones from one side of the car to the other.

The main recognition point between the BMC and Healey Motor Co Ltd anti-roll bars is the attachment to the front of the wishbones. Instead of a triangular plate, the Healey mounting used a rectangular plate with the drop link stud passing through both this and the wishbone, with the nut behind. The BMC drop link was secured to the triangular plate above the wishbone. Wishbones were not pre-drilled for the Healey anti-roll bar kit.

As with the rest of the front suspension, the anti-roll bar and fittings were painted black.

REAR SUSPENSION

The rear suspension remained unaltered until July 1962, when at chassis numbers HAN6-22907 (Sprite) and GAN1-14527 (Midget) a dust excluder was fitted to each rear spring to prevent water and road debris penetrating the cavity in and around the spring locating box. The excluder fitted over the spring in front of the rear inverted U bolt to seal the gap either side of and above the spring where it entered the rear heel panel.

In service, it had been found that this cavity could become filled with road debris and dirt, eventually causing corrosion. Initially, this could cause difficulties in releasing the two forward spring-retaining bolts as the threaded ends in the saddle clamp corroded. In extreme cases, the location box and surrounding area could become badly corroded, causing the box to move within the bulkhead, dropping the ride height as the anchor plate tilted. A point worth noting on all quarter-elliptic sprung versions is that the bottom of the anchor plate should be parallel to the bottom of the sill.

The springs were also subject to settling and giving uneven ride height from side to side. Due to customer reaction, BMC responded by introducing a tapered packing piece to place between the spring and the anchor plate, a service bulletin to dealers instructing the fitment of a packing piece rather than the replacement of a settled spring.

Special fibre washers were fitted between the rear spring eye bush and axle bracket to reduce transmitted noise into the car. These were fitted from HAN6-12068 and GAN1-7442 (December 1961), but could be added retrospectively to all earlier cars.

STEERING

The entire steering gear of the Sprite MkI was carried over into the Sprite MkII (HAN6) and Midget MkI (GAN1).

The Midget, however, used a re-styled steering wheel, with plain rounded spokes rather than the ribbed-faced spoke design of the Sprite. More noticeable, the Midget wheel was formed in a creamy white plastic as opposed to black, which continued in the Sprite. The plastic column surround always matched the colour of the steering wheel. This rather garish feature was not to last for very long: from GAN1-5035 (November 1961), the wheel was produced in black but retained the modified style which distinguished it from the Sprite wheel. The Midget also used a different centre horn push button containing an MG emblem, whereas the Sprite continued with the 'lightning flash' motif.

Certain European countries were now requiring steering locks to be fitted as standard, so Sprites and Midgets so equipped carried a combined steering lock, ignition and starter switch mounted on the right-hand side of a modified steering column.

From HAN7-28369 and GAN2-18473 (February 1963), the steering arms were revised and marked with a line of green or white paint to identify them from the superseded steering arms, the new arms carrying part numbers BTA 648 (right) and 649 (left).

BRAKES

Initially, the braking system for the HAN6 and GAN1 remained unchanged from that used in the Sprite MkI, but from chassis number HAN6-20792 and GAN1-13555 (May 1962), the rear brakes were modified to include new twin-piston slave cylinders.

The two pistons operated within the same cylinder bore to provide an equal pressure to the single trailing brake shoes. The old Micram adjuster was replaced with a cone and tappet adjuster placed opposite the slave cylinder. A square-headed adjuster on the rear face of the brake back plate could be turned to move the cone in relation to the tappets, which would then expand to move the two brake shoes an equal amount. Because the slave cylinder was now in a fixed position, the handbrake reaction lever now operated the two shoes via a separate lever placed between each shoe, the two being pivoted together. The new slave cylinder contained a bleed valve as the brass union fitted to the previous single-piston cylinder was deleted.

The basic tool kit came in a Vynide bag. This is the kit for the Sprite MkII and Midget MkI. The feeler gauges are not original to the kit – a simple two-feeler strip, for spark plugs and points, would be correct here.

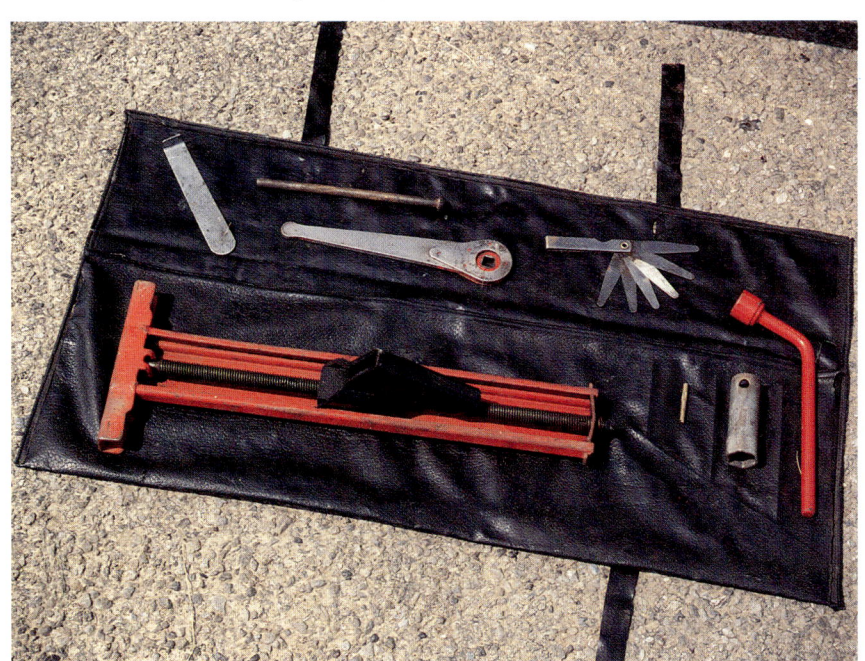

Exactly the same brake drums were used as before and they still contained the Micram adjuster hole, which continued to be blanked off with a rubber bung. At chassis numbers HAN6-21046 and GAN1-13868 (May 1962), the chrome-plated handbrake lever was shortened by 2in (50mm) and cranked slightly upwards.

With the introduction of the HAN7 and GAN2 versions in October 1962, disc front brakes became standard. This new braking, however, was neither the Dunlop/Girling nor the Lockheed 'thin disc' system which the Healey Motor Co Ltd had previously been offering, but a new Lockheed system; BMC policy dictated that all vehicles with the A-series engine would be Lockheed-braked.

Disc diameter was 8¼in (209mm) with a two-cylinder caliper placed to the rear of the wheel centre line. The caliper was secured by two $\frac{7}{16}$ UNF bolts to a new stub axle forging. A pressed steel dust shield fitted behind the disc, secured by a single $\frac{5}{16}$ UNF screw to the stub axle at its leading edge, and by the two caliper bolts at the rear. The caliper bolts served a further purpose also, for the heads were produced with a stud extension of $\frac{5}{16}$ UNF thread to secure a slotted bracket, which would ensure that the flexible brake hose would only be fitted to the caliper in the correct position – this was important as an incorrectly fitted hose could be chafed by the tyre on full steering lock.

To suit the new system, the tandem master cylinder bores were reduced from $\frac{7}{8}$in (22.23mm) diameter to $\frac{3}{4}$in (19mm) for both brake and clutch operation. Externally, the cylinder and pedal box arrangements were unchanged, except that the reservoir filler cap changed from metal to plastic at an unspecified point after the adoption of disc front brakes. From chassis numbers HAN7-30515 and GAN2-19982 (April 1963), the calipers were revised to include brake pads with a larger frictional area, to improve pad life. The new pads had a total area of 18sq in and a total swept area of 135.28sq in.

The brake and clutch pedal rubber treads now had a vertical rib pattern, the Austin motif rubbers most probably being discontinued fairly early in Sprite MkI production.

WHEELS & TYRES

Initially, both the Sprite MkII and Midget MkI continued with the same wheel and tyre specification as the Sprite MkI. The Midget, however, used a plain chromed hub cap without a logo, the Sprite continuing with the distinctive 'AH' hub cap.

In December 1962, from HAN7-26240 and GAN2-17166, the ventilated steel wheels were replaced with plain steel wheels of improved strength but remaining at 3.5in × 13in.

Centre-lock 60-spoke wire wheels of 4.0J rim × 13in became available as an alternative to disc wheels shortly after the introduction of the HAN7 and GAN2 models. Usually, the starting point of this option was given by body number rather than chassis number, these being ABL 20812 (Sprite) and GBE 16780 (Midget).

Fitting wire wheels required a number of modifications to be made. The wheels located on splined hubs with conical seating faces. The splined hubs required revised brake discs to be fitted at the front, while at the rear a slightly narrower axle casing was fitted with revised half shafts to press-fit into the centre of the splined hubs. The wheel bearing hubs were fitted with shorter studs as these were not required to secure the road wheel. The nuts securing the brake drums were locked with tab washers.

Wire wheels were secured to the hubs by chromed, two-eared, knock-on spinners which contained the outboard conical seating face. The left-hand spinners were of right-hand thread, and *vice versa*. The direction of the thread was identified on the spinner by the wording 'LEFT (NEAR) SIDE' and 'RIGHT (OFF) SIDE'. An arrow denoted the rotation to undo.

In certain export countries, the 'eared' spinner contravened local safety regulations and was replaced with a chromed octagonal-headed nut – bitter reminders of Boadicea perhaps? These nuts were not marked with '(NEAR)' and '(OFF)', but retained the other markings.

A lead mallet replaced the wheel nut brace on cars equipped with wire wheels, an octagonal spanner being an addition on cars not fitted with the 'eared' spinners.

For the steel disc wheels, an optional trim embellisher was available, known as the Ace Mercury wheel disc by Cornercroft Ltd. These appeared in three forms, and each was 'handed' to a particular side of the car to promote air flow through the wheel for brake cooling. For the Sprite, the Ace Mercury wheel disc had either an octagonal centre piece or a pseudo 'eared' spinner containing a letter S. For the Midget, a similar octagonal centre piece contained the MG logo, but no 'eared' disc was offered. A disc was held in place by a threaded adaptor pushed through the wheel centre from behind, the spinner or nut being tightened to the adaptor to retain the disc.

Tyres continued to be Dunlop C41 nylon-braced crossply type of 5.20-13, with the options of heavier duty six-ply and whitewall tyres. Steel and wire wheels were painted 'aluminium' silver grey.

IDENTIFICATION, DATING & PRODUCTION

Although they shared the same basic bodyshell unit, the Sprite MkII and Midget MkI were given individual identification prefixes for both chassis and body numbers.

PRODUCTION FIGURES

Year	RHD Home	RHD Export	LHD Export	LHD North America	RHD CKD	LHD CKD	Total
Sprite MkII (HAN6)							
1961	1607	295	8028		90	–	10020
1962	1989	205	8000		236	–	10430
Total	3596	500	3214	12814	326	–	**20450**
Sprite MkII (HAN7)							
1962	179	48	251	989	144	–	1611
1963	1400	149	1523	5064	716	–	8852
1964	31	8	9	588	116	–	752
Total	1610	205	1783	6641	976	–	**11215**
Midget MkI (GAN1)							
1961	1124	266	1353	4889	24	0	7656
1962	2295	498	1727	3832	24	48	8424
Total	3419	764	3080	8721	48	48	**16080**
Midget MkI (GAN2)							
1962	288	114	384	678	6	12	1482
1963	2066	313	1654	3562	30	0	7625
1964	140	10	1	337	6	0	494
Total	2494	437	2039	4577	42	12	**9601**

The Sprite was identified by the addition of an H for Healey, so the chassis number prefix became HAN6, the 6 being the sixth in the series of Austin-Healey models. The number sequence did not follow on from the AN5, but started afresh from 101. The Midget was given the prefix GAN1, G being the code for MG and 1 denoting that this was the first in the series of MG Midgets. Again, the numbering sequence began at 101.

The chassis number plate was located in the same position as before but a new plate was provided for the Midget. This contained the MG octagon with the wording 'Made in England' in the top left-hand corner, and 'When Ordering Replacements Quote' in the top right-hand corner. Below this, the spaces for car (chassis) number and engine number were the same as those on the Sprite chassis plate. Again, the space for the engine number was stamped 'See Engine'. As before, left-hand drive cars were identified with a letter L preceding the actual chassis number.

The body number, applied at the Swindon factory, was carried on a screwed-on plate in the left-hand door hinge pillar, the prefixes being ABL (Sprite) and GBE (Midget). This plate was sprayed over in body colour, whereas the chassis number plate was fitted at the Abingdon assembly plant after the bodyshell had

While the Sprite MkII continued with the same chassis number plate as the 'Frogeye', the Midget used a more appropriate plate with an MG logo. The numbers should read from the left-hand side of the car. As before, the engine number was not included on this plate.

been sprayed at Cowley. The right-hand door pillar carries a similar plate with a number prefixed by BAL (Sprite) or BGE (Midget). These secondary body numbers are not usually listed in the production records.

Australian-assembled Sprites, available from August 1962, became YHAN-6, Midgets not being supplied in CKD form to Australia. The manufacturing period code numbering system used for the Sprite MkI was discontinued at the end of November 1961, so this no longer appeared between the prefix and the actual chassis number.

For the German market, an additional identification plate was fixed on the bonnet slam panel to the right-hand side. In addition to the chassis number, this plate also contained the year of manufacture, the total load, and maximum loads for front and rear axles. The chassis number was also stamped into the right-hand inner wing panel near the brake pipe brass union.

The engine number for Sprite HAN6 and Midget GAN1 models began 9CG-DA-H (High compression) or L (Low compression). The DA referred to the close-ratio gearbox now fitted as standard. The numbering sequence began at 101, with both Sprites and Midgets drawing engines from the same sequence. Australian-assembled Sprites continued to carry the engine number on the bulkhead-mounted chassis number plate in addition to colour of the bodywork.

Cars assembled from CKD kits used chassis and engine numbers from the home market sequence of alloted numbers. Semi-CKD kits bound for Milan to be built as Innocenti versions continued to use home market chassis numbers but had engine number prefixes as follows: Innocenti Spyder, up to 1963, 948cc, 9CB; Innocenti 'S', 1963 onwards, 1098cc, 10CB.

From October 1962, the prefix for Sprites and Midgets became HAN7 and GAN2 respectively, this change marking the arrival of disc front brakes, a 1098cc engine, an improved gearbox and revised interior trim. The numbering sequence continued unchanged, and indeed would remain so for the entire production run in the years ahead. The change of prefix occurred at HAN7-24732 and GAN2-16184, indicating that the Sprite was comfortably outselling the Midget.

The increase in engine size and prefix did not bring about a change in 'Mk' designation for either Sprite or Midget at this stage, but the Australian 1098cc Sprite, YHAN7 (available from March 1963), became known locally as the MkIIA to distinguish it from the 948cc MkII. The new engine number sequence, prefixed 10CG-DA-H or L, began at 101.

As with the Sprite MkI, the relationship between chassis, body and engine numbers was somewhat haphazard, and best explained by the assembly procedure. Body numbers were applied at Swindon before the bodyshells were taken to Cowley for painting and then on to Abingdon for assembly. This movement and storage of bodyshells from one factory to another would disturb the numerical sequencing of body numbers before arrival on the assembly line. This could also be compounded by bringing required colours onto the line quickly, leaving some bodies to sit for a period until their colours were required.

Chassis number plates were fitted on the assembly line, but the sequencing of these could also be disturbed, because there was more than one assembly line. Engines were supplied already numbered from the Morris Engine Plant in Coventry, then drawn randomly from the Abingdon stores to be fitted into a Sprite or a Midget regardless of chassis or body number.

With the sourcing of parts in this way, it is easy to see why the numerical progression of chassis, body and engine numbers was not constant. This situation remained throughout the production life of all Sprites and Midgets.

OPTIONS, EXTRAS & ACCESSORIES

The following items were available for HAN6/7 and GAN1/2 models:
Radio
Heater

PRODUCTION DATING

Date	Sprite	Midget	Notes
Feb 61	HAN6-101	–	Announced 21 May
Mar 61	–	GAN1-101	Announced 20 Jun
Jan 62	HAN6-12247	GAN1-7526	First cars built in 1962
Oct 62	HAN6-24731	GAN1-16183	Last HAN6/GAN1 cars
Oct 62	HAN7-24732	GAN2-16184	1098cc engine and disc front brakes
Jan 63	HAN7-26622	GAN2-17688	First cars built in 1963
Jan 64	HAN7-37735	GAN2-25361	First cars built in 1964
Mar 64	HAN7-38828	GAN2-25787	Last HAN7/GAN2 cars

The BMC hardtop that was commonised for both Sprite MkII and Midget MkI. Prior to this, the Sprite used a slightly revised version of that offered for the Sprite MkI. The Midget, however, used a similar pattern to the one shown, but with a back-sloping rear edge for the sidescreen, necessitating different sidescreens for the hood and hardtop. Strictly speaking, the hardtop on a Deep Pink car would have been Grey or Old English White.

Fresh air unit
Tonneau cover and rail
Laminated windscreen
Locking fuel filler cap
Cigar lighter
Wing-mounted mirror
Ace Mercury wheel discs
Hardtop
Whitewall tyres
Wire wheels HAN7 and GAN2 only
Heavy duty six-ply tyres
Twin Windtone horns
Luggage carrier
Low 8.3:1 compression ratio
Seat belts Dealer fitted, usually Kangol
Rear compartment cushion
Front anti-roll bar From January 1964
Supplementary tool kit in a roll (Part number 97H 524) containing four spanners, 6in pliers, 7in adjustable spanner, ⅜in diameter tommy bar, tubular spanner (½ × ⁹⁄₁₆ AF) and Philips screwdriver.

A windscreen washer and cable-driven rev counter were options on the rare 'basic' model, but standard on the de luxe.

BMC also offered a Lockheed brake servo kit for several of its vehicles, including Sprites and Midgets. However, this did not seem to appear on any of the accessory or special tuning listings dedicated to the Sprite and Midget. The kit was part numbered 8G 8732 and dated to August 1963 on the approved accessories salesman's guide.

As described earlier in the 'Weather Equipment' section, the hardtop option initially differed between Sprite and Midget, the latter requiring special sidescreens with a rearwards-sloping trailing edge. The Sprite hardtop was very similar to that of the MkI, whereas the Midget used a more angular version with a flat rear screen. The Sprite hardtop, available only in Old English White, carried part number AHA 6193. The Midget hardtop was available in four colours, part numbers being AHA 6301 (Red), AHA 6302 (Blue), AHA 6303 (Grey) and AHA 6323 (Old English White). The special sidescreens for the Midget hardtop had part numbers AHA 6150 (right) and AHA 6151 (left).

Later, the optional hardtop was commonised to both Sprite and Midget. This was based on the Midget item but now used the normal sidescreens. Part numbers were AHA 6458 (primed finish), AHA 6459 (Red), AHA 6460 (Blue), AHA 6461 (Grey)

Healey offered its own hardtop for the Sprite MkII which, as can be seen here, obviously fitted the Midget MkI just as well. Standard BMC sidescreens were now used, and both panes could now slide in the frame.

and AHA 6462 (Old English White).

To spoil – or confuse! – an owner even further, the Healey Motor Co Ltd offered its own restyled hardtop. This contained rear quarter windows to give a lighter and more airy appearance than BMC's rather van-like design. This also now used the standard pattern sidescreen.

Healey continued to offer its own range of accessories and speed equipment, but when wire wheels and disc brakes became available from the factory on the 1098cc versions, the Healey conversion kit was no longer offered. The supercharger kit also disappeared, no doubt as a result of overheated engines and too many complaints from overheated owners. A carpet set in red, blue or black was also offered by Healey to replace the rubber matting of the earlier versions.

BMC Special Tuning no longer produced a booklet to describe its offerings and advice. Instead, loose-leaf lithographed sheets were regularly revised as further development and experience improved the product range. The Sprite HAN6/7 and Midget GAN1/2 were catered for respectively by publications C-AKD 1021 (a later loose-leaf issue than the Sprite MkI booklet bearing the same number) and C-AKD 5097. Engine tuning equipment was not offered for the 1098cc 10CG-type engine owing to the crankshaft rigidity problem. It would take too much space to describe all the special tuning parts from all the issues, but the following list summarises the main items that were available, with part numbers:

Bonnet securing straps C-HHH 5517-8-9
DS11 brake pad set (HAN7, GAN2) C-AHT 16
S1 camshaft 2A 948 88G 229

The front of the glassfibre hardtop was secured to the top of the windscreen frame with two clamps which bolted through the glassfibre.

Competition camshaft C-AEA 731
Distributor for use with above C-27H 7766
Twin 1½in SU carburettor assembly C-AUD 194
Installation kit for above C-AJJ 3304
Intake flare pipes for above (steel) C-AEA 485
Intake flare pipes for above (fibreglass) C-AHT 10
Intake flare pipes for above (alloy) C-AHT 247
Competition crankshaft for 948cc engines 9C and 9CG, coded 'Red Type' C-AEA 406
Large-bore exhaust manifold C-AHT 11
Large-bore exhaust pipe C-AHA 5449
Silencer for above C-ARA 135
Competition clutch assembly Up to engine number 9CG-2139, C-AEJ 31; from engine number 9CG-2140, C-AEJ 77
Deep sump and oil pick-up kit C-AJJ 3324
Duplex timing gear set C-AJJ 3325

Notes to table
[1] Fleck carpet covered the axle tunnel and rear wheelarches.
[2] Rubber matting on the floor did not extend under the seats on HAN6 models.
[3] Bright Red trim coverings replace Cherry Red from 14842 to 14965, 14980 to 15211, 15218 onwards (Feb 62).
[4] Black with Red Fleck replaced by Red with Black Fleck from 16472 (Mar 62).
[5] Iris Blue replaces Speedwell Blue from 5133 (Sep 61).
[6] Blue trim changed to the same shade used in the Midget, from 6285 (Sep 61).
[7] Early hardtops were Old English White; colour options became available with the later, flat rear window type commonised with the Midget.

COLOUR SCHEMES (HAN6 SPRITE MKII)

Body colour (paint code)	Seats	Seat piping	Trim	Carpet[1]	Matting[2]	Hood	Tonneau	Optional hardtop[7]
Black (BK1)	Cherry Red[3]	Black	Cherry Red[3]	Red/Black Fleck	Cherry Red	Grey	Red	Grey or Old English White
	Red[3]	Black	Red[3]	Red/Black Fleck	Cherry Red	Grey	Red	Grey or Old English White
Iris Blue (BU12)[5]	Blue[6]	Blue[6]	Blue[6]	Blue/Black Fleck	Blue	Blue	Blue	Blue or Old English White
Speedwell Blue (BU1)[5]	Blue	Blue	Blue	Blue/Black Fleck	Blue	Blue	Blue	Blue or Old English White
Deep Pink (RD18)	Black	White	Black	Black/White Fleck	Black	Black	Black	Grey or Old English White
Signal Red (RD2)	Black	Red	Black	Black/Red Fleck[4]	Black	Black	Black	Red or Old English White
	Cherry Red[3]	Black	Cherry Red[3]	Red/Black Fleck	Cherry Red	Black	Red	Red or Old English White
	Red[3]	Black	Red[3]	Red/Black Fleck	Cherry Red	Black	Red	Red or Old English White
Old English White (WT3)	Black	White	Black	Black/White Fleck	Black	Grey	Black	Red, Grey, Blue or Old English White
	Cherry Red[3]	White	Cherry Red[3]	Red/Black Fleck	Cherry Red	Grey	Red	Red, Grey, Blue or Old English White
	Red[3]	White	Red[3]	Red/Black Fleck	Cherry Red	Grey	Red	Red, Grey, Blue or Old English White
Highway Yellow (YL9)	Black	White	Black	Black/White Fleck	Black	Black	Black	Grey or Old English White

ORIGINAL SPRITE & MIDGET

High compression 9.3:1 pistons Available for standard, +.010in, +.020in, +.030in and +.040in bore sizes, C-2A 946
Large capacity oil pump 12G 793
Competition front shock absorbers C-AHA 6451-2
Adjustable rear shock absorbers C-AHA 6453-4
Front suspension lowering kit C-AJJ 3322
Anti-roll bar kit C-AJJ 3314
Straight-cut close-ratio gear set C-AJJ 3319
Half shaft for disc wheel C-BTA 940
3.727:1 crown wheel and pinion* BTA 535
4.55:1 ratio crown wheel and pinion* C-BTA 816
Limited slip differential C-BTA 881 or 696
Wire wheels 60-spoke, 5in × 13in (HAN7, GAN2), C-AHA 7573

Note
*The differential carrier was changed upon the introduction of the 1098cc engine, and only allowed the fitting of two alternative ratios from standard. Prior to this, all of the alternative ratios offered for the MkI Sprite could still be fitted.

COLOUR SCHEMES

All details of colour combinations are given in the four accompanying tables. Irrespective of body colour, the area behind the radiator grille was sprayed black.

Australia continued to use locally applied colour schemes. The exact colour specification seemed to depend upon which supplier of paint was used. In all cases, the colour, manufacturer and type of paint was given on the bulkhead chassis number plate.

PRODUCTION CHANGES

CHANGES BY CHASSIS NUMBER

HAN6-101 (Feb 61), GAN1-101 (Mar 61)
First chassis numbers allocated: Sprite MkII announced 21 May, Midget MkI announced 20 June.
HAN6-3548 (Jun 61), GAN1-1209 (Jul 61)
Improved boot hinges.
HAN6-3808, GAN1-1043 (Jul 61)
Rev counter cable lengthened by 1in to 26in in RHD cars only.
GAN1-1508 (Jul 61)
An additional plate stud to secure Midget bonnet moulding chrome strip.
HAN6-5660, GAN1-2206 (Aug 61)
Changed windscreen wiper motor on LHD cars.
HAN6-6276, GAN1-3007 (Sep 61)
Changed windscreen wiper motor on RHD cars.
HAN6-6285 (Sep 61)
Changed blue trimming material for Sprite to match that fitted to Midget.
HAN6-6723 (Sep 61)
Black trim covering revised.
HAN6-7499, GAN1-3491 (Sep 61) intermittently to HAN6-10118, GAN1-7520 (Dec 61)
Revised dynamo and control box.

The underside of the Midget boot lid shows the multitude of spire clips and sealant used to keep the badging in place. Boot lids tend not to rust so vigorously as bonnets, but this is not to suggest that they do not rust at all!

Notes to table
[1] Use of carpet and matting on GAN1 Midgets: fleck matting covered the entire floor; fleck carpet covered the axle tunnel, rear wheelarches and lower rear of seat squabs.
[2] From chassis number GAN1-10742 (Mar 62), rear of seat squabs and rear axle tunnel carpeting changed from Black/Red Fleck to Red/Black Fleck.
[3] Clipper Blue replaced by Ice Blue from GAN1-2854 (Sep 61).
[4] Almond Green used on 12 home market Midgets, GAN1-13376 to 13387 (Apr 62) plus GAN1-8362 (Jan 62).
[5] Farina Grey replaced by Dove Grey from GAN1-2756 (Sep 61).

COLOUR SCHEMES (GAN1 MIDGET MKI)

Body colour (paint code)	Seats	Seat piping	Trim	Carpet and matting[1]	Hood	Tonneau	Optional hardtop
Black (BK1)	Red	Black	Red	Black/Red Fleck[2]	Grey	Red	Grey or Old English White
Clipper Blue (BU14)[3]	Blue	Blue	Blue	Blue/Black Fleck	Blue	Blue	Blue or Old English White
Ice Blue (BU18)[3]	Blue	Blue	Blue	Blue/Black Fleck	Blue	Blue	Blue or Old English White
Almond Green (GN37)[4]	Green	Green	Green	Black/White Fleck	Grey	Black	Old English White
Dove Grey (GR26)[5]	Red	White	Red	Red/White Fleck	Grey	Red	Grey or Old English White
Farina Grey (GR11)[5]	Red	White	Red	Red/White Fleck	Grey	Red	Grey or Old English White
Tartan Red (RD9)	Red Black	Black Red	Black Red	Red/Black Fleck Black/Red Fleck[2]	Red Red	Black Red	Red or Old English White Red or Old English White
Old English White (WT3)	Black Red	White White	Black Red	Black/White Fleck Red/White Fleck	Grey Grey	Black Red	Grey or Old English White Grey or Old English White

SPRITE MKII & MIDGET MKI

Rubber floor matting was still used for the Sprite (HAN6) and Midget (GAN1) despite the use of carpeting behind the seats on the axle tunnel and wheelarches. Despite what is said elsewhere about the poor durability of rubber matting, here is an example which has outlasted the car!

Note to table
[1] Change from British Racing Green GN25 to GN29 occurred from body number 27492 (Aug 63).

HAN6-7960, GAN1-3734 (Oct 61)
Battery tray sides changed from plain to recessed.
HAN6-8157, GAN1-4103 (Oct 61)
Four rubber buffer blocks fitted to the boot drip channel to improve support for the boot lid.
HAN6-8692, GAN1-4582 (Oct 61)
Boot handle changed from 'L' to 'T' pattern.
HAN6-9476, GAN1-4804 (Oct 61)
New headlamp and rim for LHD cars, except Europe and USA.
GAN1-5034 (Nov 61)
White steering wheel and column surround deleted.
HAN6-10686, GAN1-5892 (Nov 61)
New headlamp and rim for European LHD cars, except France and Sweden.
HAN6-10720 (Nov 61), GAN1-6021 (Dec 61)
Revised throttle pedal fitted.
HAN6-10913, GAN1-6167 (Nov 61)
Oil pressure gauge sender pipe flexible hose changed.
HAN6-11162, GAN1-6255 (Dec 61)
Rev counter drive gearbox changed. Large brass nut replaced with a knurled nut to prevent overtightening.
HAN6-11691, GAN1-6871 (Dec 61)
New headlamp and rim for France.

COLOUR SCHEMES (HAN7 SPRITE MKII)

Body colour (paint code)	Seats	Seat piping	Trim	Carpet and matting[1]	Hood	Tonneau	Optional hardtop
Black (BK1)	Red Hazelnut	Grey Grey	Red Hazelnut	Cardinal Red Hazelnut	Black Hazelnut	Black Hazelnut	Grey or Old English White Grey or Old English White
Iris Blue (BU12)	Blue	Grey	Blue	Blue	Blue	Blue	Blue or Old English White
British Racing Green (GN25 or 29)[1]	Black	Grey	Black	Black	Black	Black	Old English White
Dove Grey (GR26)	Red	Grey	Red	Cardinal Red	Grey	Red	Grey or Old English White
Signal Red (RD2)	Red Black	Grey Grey	Red Black	Cardinal Red Black	Red Red	Red Red	Red or Old English White Red or Old English White
Old English White (WT3)	Hazelnut Red Black	Grey Grey Grey	Hazelnut Red Black	Hazelnut Cardinal Red Black	Hazelnut Grey Black	Hazelnut Red Black	Old English White Old English White Old English White
Fiesta Yellow (YL11)	Black	Grey	Black	Black	Black	Black	Old English White

HAN6-11703, GAN1-6679 (Dec 61)
  Oil pressure gauge sender pipe between engine and flexible hose changed.
HAN6-11769, GAN1-6738 (Dec 61)
  Headlamp rims and sealing rubber gasket changed.
HAN6-11970, GAN1-6644 (Dec 61)
  Blanking rubber bung for clutch slave cylinder changed. Vertical web added to the horizontal web.
GAN1-7702 (Jan 62)
  MG octagon boot lid badge revised.
HAN6-12068, GAN1-7442 (Dec 61)
  Special fibre washer added to rear spring to axle bolt to reduce noise.
HAN6-12717, GAN1-8991 (Jan 62)
  Boot lid sealing rubber changed for Switzerland.
HAN6-12777, GAN1-7681 (Jan 62)
  New headlamp and rim for USA.
HAN6-12868, GAN1-7898 (Jan 62)
  Rear brake linings changed.
HAN6-15219, plus 14842-14965 and 14980-15211 (Feb 62)
  Trim coverings changed from cherry red to bright red.
GAN1-10742 (Feb 62)
  Revision to interior trim colour combination: see 'Colour Schemes' section.
HAN6-16472 (Mar 62)
  Rear axle tunnel carpet changed from black with red fleck to red with black fleck for some colour schemes.
HAN6-16651, GAN1-10742 (Mar 62)
  Bonnet release cable clamp end changed.
HAN6-17637, GAN1-11890 (Mar 62)
  Boot lid sealing rubber changed.
HAN6-20093, GAN1-13404 (Apr 62)
  Brake and clutch pedals changed.
HAN6-20105, GAN1-13449 (May 62)
  Windscreen washer kit and control knob changed.
HAN6-20792, GAN1-13555 (May 62)
  Rear brakes changed, slave cylinders now with two pistons. Brake adjustment now effected via a square-headed screw from behind the backplate.
HAN6-20820, GAN1-13763 (May 62)
  Thicker diameter front bump stop rubber buffers fitted.
HAN6-21046, GAN1-13868 (May 62)
  Handbrake lever shortened and cranked upwards.
HAN6-22907, GAN1-14527 (Jul 62)
  Dust excluders fitted to rear springs to seal off spring location boxes.
HAN6-23843, GAN1-15309 (Aug 62)
  Reinforcement plate fitted behind door check strap to prevent distortion to the edge of the A post.
HAN6-24125, GAN1-15850 (Sep 62)
  Towing eyes fitted to front frame rails (Canada).
HAN7-24732, GAN2-16184 (Sep 62)
  Introduction of 1098cc engine, baulk ring synchromesh gearbox, disc front brakes and revised interior trim.
HAN7-26240, GAN2-17166 (Dec 62)
  Ventilated steel disc road wheels replaced with plain disc wheels of greater strength. Rear brake cylinders now located by roll pins in modified backplates.
HAN7-27756, GAN2-18220 (Jan 63)
  Battery retaining bar and rods changed to suit new nine-plate-per-cell battery (previously seven).
HAN7-28148, GAN2-18476 (Feb 63)
  Rear-view mirror changed.
HAN7-28369, GAN2-18473 (Feb 63)
  Improved steering arms fitted.
HAN7-30515, GAN2-19982 (Apr 63)
  Brake calipers and disc pads changed. Pad contact area increased from 3.3sq in to 4.14sq in (8.38 to 10.5sq cm).
HAN7-31011, GAN2-20208 (Apr 63)
  Sliding pane sidescreen frames now fitted with compression springs and thinner felt strips to prevent rattling and improve sliding action.
HAN7-31701, GAN2-20941 (May 63)
  Hazelnut seat cushion changed.
HAN7-32076, GAN2-21041 (May 63)
  Black seat cushion changed.
HAN7-32095, GAN2-21091 (May 63)
  Red seat cushion changed.
HAN7-32148, GAN2-20886 (Jun 63)
  Blue seat cushion changed.
HAN7-32388, GAN2-20933 (Jun 63)
  Jack and ratchet handle changed.
HAN7-38828, GAN2-25787 (Jan 64)
  Last chassis numbers allocated.

Notes to table
[1] Change from British Racing Green GN25 to GN29 occurred from body number 22599 (Aug 63).
[2] Hazelnut trim and hood was rare with British Racing Green.

COLOUR SCHEMES (GAN2 MIDGET MKI)

Body colour (paint code)	Seats	Seat piping	Trim	Carpet and matting[1]	Hood	Tonneau	Optional hardtop
Black (BK1)	Red	Grey	Red	Red	Black	Black	Grey or Old English White
Ice Blue (BU18)	Blue	Grey	Blue	Blue	Blue	Blue	Blue or Old English White
British Racing Green (GN25 or 29)[1]	Black Hazelnut[2]	Grey Grey	Black Hazelnut[2]	Black Hazelnut[2]	Black Hazelnut[2]	Black Hazelnut[2]	Old English White Old English White
Dove Grey (GR26)	Red	Grey	Red	Cardinal Red	Grey	Red	Grey or Old English White
Tartan Red (RD9)	Red Black	Grey Grey	Red Black	Cardinal Red Black	Red Red	Red Red	Red or Old English White Red or Old English Whiteß
Old English White (WT3)	Hazelnut Red Black	Grey Grey Grey	Hazelnut Red Black	Hazelnut Cardinal Red Black	Hazelnut Grey Black	Hazelnut Red Black	Old English White Old English White Old English White

SPRITE MKII & MIDGET MKI

These two nuts served the upper seat belt mountings, although seat belts were a listed option only. This close view also shows the flecked carpeting and Vynide-covered cockpit surround.

CHANGES BY ENGINE NUMBER

9CG 101 (Feb 61)
First engine.

9CG 1154 (Apr 61)
Dynamo adjusting link improved and larger diameter washer fitted.

9CG 2140 (Jun 61)
Gearbox casing revised with improved clutch fork lever and clutch cover unit, thicker clutch pressure plate.

9CG 3170 (Jun 61)
Inlet manifold balance pipe core plugs revised with dished plugs in enlarged manifold counter bores, replacing convex Welch type plugs.

9CG 4523 (Jun 61)
Improved exhaust valve material to prolong life.

9CG 5067 (Jul 61)
Oil filter changed to vertically hung mounting.

9CG 5093 (Jul 61)
Crankshaft timing gear changed from cast iron to steel.

9CG H8985 (plus 8690-8700), 9CG L1043 (Sep 61)
Distributor rotated in housing necessitating lengths of HT leads and vacuum advance pipe to be suitably altered.

9CG H12256, L10996 (Nov 61)
Improved rocker shaft.

9CG H15258, 9CG L14245 (Dec 61)
Burman vane type oil pump no longer fitted.

9CG 22556 (Feb 62)
Aluminium rocker shaft pillars replaced with cast iron.

9CG H22851, 9CG L15656 (Feb 62)
Fan belt changed.

9CG H27445, 9CG L16949 (Mar 62)
Distributor revised.

9CG 32074 (Jun 62)
Improved carburettor float chamber needle valve assemblies to obviate flooding.

9CG H32735, 9CG L26633 (Jun 62)
Water pump revised.

9CG 36711 (Oct 62)
Last 948cc engine.

10CG 101 (Oct 62)
First 1098cc engine.

10CG 2572 (Nov 62)
Dynamo adjusting link pillar stud lengthened and dynamo top securing bolts to suit revised dynamo.

10CG H5614, L1945
Revised pistons and connecting rods with gudgeon pins of reduced bore size.

10CG 6868 (Mar 63)
Improved front and rear tappet chest covers introduced to counter oil leakage problem.

10CG 10345 (May 63)
New clutch cover unit with stronger springs.

10CG 14878 or 18628 (see engine section) (Aug 63/Nov 63)
Inlet valves increased from 29.4mm to 31mm head face diameter.

10CG 15607
Thermostat changed from downwards to upwards-opening.

10CG 17773 (Oct 63)
Fan belt changed.

10CG 18628 (Nov 63)
Inlet manifold fitted with location spigots to cylinder head, larger inlet valves (31mm) and reshaped combustion chambers.

10CG 21048 (Jan 64)
Last 10CG engine.

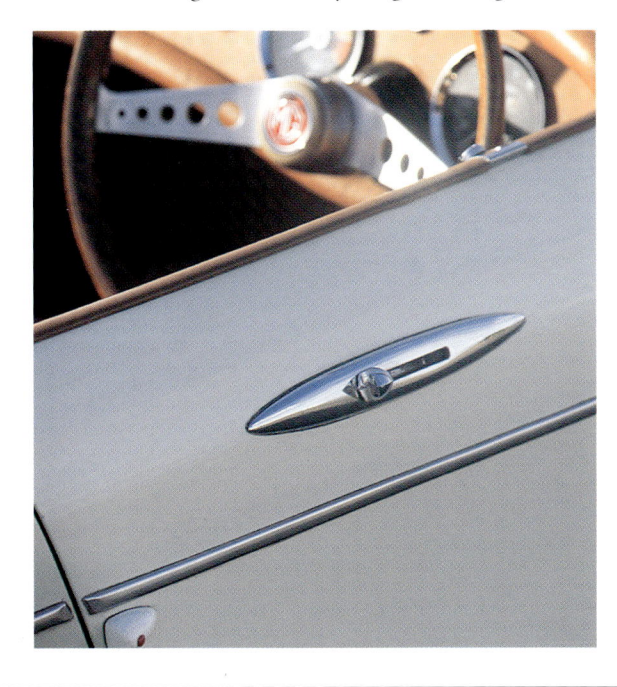

Not original, but the Speedwell exterior door handle is an interesting and popular period piece. The steering wheel is also non-original and must represent the most popular item of all to change on the early Spridgets. Speedwell produced a vast range of accessories and modifications for the Sprite and Midget.

SPRITE MkIII & MIDGET MkII
(HAN8, GAN3)

By now BMC now no longer had the small sports car market to itself, for Triumph had launched its 1147cc Spitfire in 1962. Based on the Herald, the Spitfire offered a number of features which the 'Spridgets' lacked.

BMC and Healey addressed this in March 1964 with the Sprite MkIII (HAN8) and Midget MkII (GAN3). Although these were very much the same in overall appearance as the earlier models, new doors with winding windows, exterior door handles and locks provided a balancing of features against the Spitfire. Minus points, perhaps, were the loss of door pockets and elbow room, the space now being taken by the window winding mechanism. A new dashboard, instrument and control layout was introduced, along with a small parcel tray on the passenger side to help make up for the loss of the door pockets. A new windscreen, side quarterlights and an improved hood added comfort, but a heater was still optional.

Remaining at 1098cc, the engine was substantially improved as the earlier version had suffered from a crankshaft rigidity problem. Enlarged main bearing diameters, an improved cylinder head and a different exhaust system saw power output increase to 50bhp. The new engine had no provision for a camshaft-driven mechanical fuel pump, so an SU electric pump was fitted above the right-hand rear shock absorber.

The quarter-elliptic rear springs were now discarded in favour of more conventional semi-elliptic leaf springs. These provided improved ride quality and lateral location, if not such good longitudinal location. Modifications required to the bodyshell to incorporate the new springs were fairly simple and, as it turned out, no more expensive to produce than a complete suspension system.

BODY & CHASSIS

Although very similar in appearance, the bodyshell for the Sprite MkIII and Midget II contained some significant changes.

Completely new door assemblies – containing a winding window, external door handle, locking barrel and a swivelling quarter light assembly – required new A- and B-posts to be fitted. The top of the A-post was reshaped where it joined the scuttle, and, accompanying this, the method of fitting the windscreen frame was changed. This now fitted into a slot with the fasteners being inside the door pillar, and shims were inserted as required between the mounting face of the frame and the door pillar. The revised B-post provided the location for a new catch plate. At the top of the B-post, the rear wing was revised in sympathy with the new shape of the top edge of the door.

The change from quarter-elliptic to semi-elliptic rear springs necessitated a revised front location box in the bulkhead. This became a simpler construction as the radius arm bracket and bottom spring anchor plate were no longer required. The spring was fitted to a plate which was secured beneath the floor by four ⅜ UNF bolts. The spring was accommodated in a recess in the floor and bulkhead, internal stiffeners rising either side to feed the stress into the inner and outer heel panels. The stiffeners also had threaded blocks welded to the sides, these accommodating the two rear bolts securing the spring plate. The front two bolts engaged with welded nuts in the plate, the bolts fitting from inside the cockpit just in front of the heel panel. The longitudinal top hat stiffener running from the heel panel, along the floor and to the jacking point crossmember was deleted, since the semi-elliptic spring considerably reduced the loadings fed into the rear bulkhead and floor.

The rear of the spring was located in a pivoting shackle, the bracket for this being bolted to the boot floor by three ⅚ UNF Nyloc nuts and bolts. Inside the boot, a strengthening bracket was welded to the side boxed member and floor to accept the loadings imposed by the shackle.

The transmission tunnel now contained welded nuts for the inner seat belt mountings.

It could be the mid-1960s, except for the price of petrol! Jonathan Whitehouse-Bird's Tartan Red Midget MkII in superb unrestored and original condition.

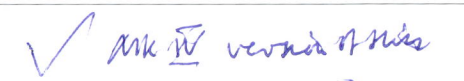

Barry Knight's Sprite shows the new windscreen and frame introduced for the Sprite MkIII and Midget MkII, but otherwise the familiar appearance was unaltered.

Sprite MkIII with optional luggage rack and wire wheels, the latter seen here in non-original chrome rather than silver paint. Note the correct 'crimped' end to the silencer box and re-aligned exit.

Sprite MkIII & Midget MkII

BODY TRIM

With the introduction of the new doors, external door handles were now fitted, made by Wilmot Breeden and of chrome-plated Mazak. Below each handle, a key-operated locking barrel was fitted, also made by Wilmot Breeden.

A pivoting quarterlight assembly was provided with a locking latch lever. The rear edge of the chrome quarterlight frame incorporated the winding window guide channel.

The windscreen frame, containing new glass of slightly increased area and curvature, now fitted into slots in the top of the A-posts. Still retained by two fasteners per side, these were now secured from the inside to delete the external screw heads of the earlier version. Shims were fitted between the base of the frame pillars where these abutted the A-post. The pillars were now a part of the frame, so retaining screws were no longer required to secure the pillars to separate frame side members. A chrome-plated tie rod was provided at the centre of the windscreen to link the top of the frame to the top of the scuttle, and also served as a mounting for the rear-view mirror, which could be adjusted for height on the tie rod. Longer 9in (228.6mm) wiper blades accompanied the new windscreen.

From Midget GAN3-49679 (June 1966), the construction of the radiator grille was revised, with an accompanying change to the MG badge. From chassis number HAN8-58701 (February 1965) and GAN3-46562 (March 1965), the single-piece front

Original Sprite & Midget

number plate mounting bracket was replaced with a simpler two-piece type for right-hand drive cars only. Changes for other markets are listed in 'Production Changes'.

The Midget continued to retain the extra external trim detailing not found on the Sprite. From August 1965, however, Australian-assembled Sprites adopted the same waistline trim mouldings as the Midget, but not the bonnet moulding. This addition on Australian Sprites coincided with wire wheels becoming standard fitment, the revisions leading the car to be designated Sprite MkIIIA.

The bonnet prop was now fitted to the right-hand side with the top of the prop pivoting from a point nearer to the centre line of the bonnet. This latter feature countered the tendency for the bonnet to sag when raised. The boot support stay rod was now finished in nickel plate, replacing the black-painted item used previously. On the inside of the boot lid, the badge retaining clips were replaced with plastic caps from HAN8-62602 and GAN3-50340 (June 1966).

WEATHER EQUIPMENT

The hood remained similar to the one used on the previous 'Spridget' versions, but was now attached to the top screen rail by two chrome-plated over-centre clips operated from within the car. The clips hung from a header rail, painted Cumulus Grey and permanently fixed to the front of the hood, and engaged with small brackets riveted to the inside of the windscreen frame rail.

The hood frame, still painted Cumulus Grey, no longer contained a spring-tensioning portion, but still consisted of two main bows that could be split in the middle for storage in the boot. The geometry of the links and pivots between the bows was altered. The rear of the hood attached to the body in the same fashion as before.

The optional tonneau cover now contained zips running back from the B-posts to allow the cover to fold down behind the seats over the support rail, the cover fastening to the heel panel pegs as before. The support rail, which split in the centre for storage, was also painted Cumulus Grey. The front of the cover was now provided with four 'lift the dot' fasteners, so commonising the position of the scuttle-mounted pegs for both right- and left-hand drive. This did not, of course, commonise the tonneau cover, for the steering wheel pocket still had to be on the appropriate side. To the front outside corners of the cover, additional anchorage to the base of each screen pillar was provided by a stud and button fastening.

The optional hardtop, available in Riviera Blue, Tartan Red, Black, Old English White or Primer, was completely new and common to both Sprite and Midget. It now contained rear quarter windows and utilised the same attachment points on the body and screen rail as the hood. All windows were toughened glass, set in moulded rubber seals, and provided with aluminium guttering over the windows. Since it was made of double skinned fibreglass, the hardtop had a textured inner surface rather than a stark fibreglass matting finish.

The new windscreen and frame for the Sprite MkIII and Midget MkII, showing the centre tie rod which provided height adjustment for the rear-view mirror. Wiper blades were now 1in (25.4mm) longer for the new screen.

Windscreen pillars were now secured from the inside of the A-post. Shims were used as required between the pillar and A-post to prevent stressing the pillars through any discrepancies between the two.

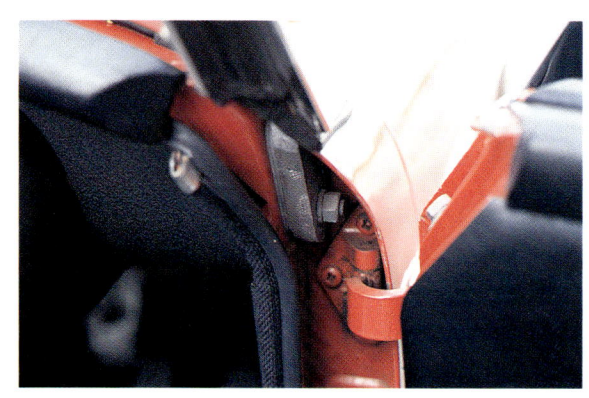

Wire wheels had to be stowed outer face upwards due to the amount of off-set of the hub. Steel disc wheels were always stowed outer face downwards to take advantage of the extra room offered by the deeply dished rear face. The boot lid prop should be plated instead of painted.

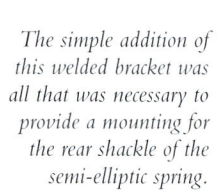

The simple addition of this welded bracket was all that was necessary to provide a mounting for the rear shackle of the semi-elliptic spring.

All present and correctly stowed in the boot of this Midget MkII. The hood bag is strapped to the cockpit divider panel, with the tonneau stick bag below, tool kit to the left and tonneau cover on top of the spare wheel. The boot lid supporting prop now had a plated rather than black-painted finish. Pads on top of the wheelarches were provided to stop the hood header rail from damaging the paintwork when stowed.

INTERIOR TRIM

The seats were initially carried over from the earlier HAN7 and GAN2 versions in coverings of black, red or a new light blue, but a completely new type of seat was introduced in August 1965 according to colour as follows: black, HAN8-55373 and GAN3-42995; red, HAN8-55625 and GAN3-42827; light blue, HAN8-55537 and GAN3-43171.

The new seat comprised a tubular-framed base unit and hinged squab frame, with the seat cushion no longer removable from the base frame. A rubber diaphragm was held to the frame with 20 hooks, with the foam cushion pad and Vynide covering assembled to this as a complete unit. The tubular-framed squab was also fitted with a foam pad to give shape and support to the covering. Fluted centre sections with plain bordering contained no transverse pleating, but the seat edging was still given piping, now in pale grey for each colour option.

The hinged backrest was provided with rake adjustment, by means of two bolts with locking nuts which could be screwed in or out of welded nuts in the base of the backrest. The heads of the bolts acted as the feet of the backrest where this met the lower part of the seat frame assembly. Both driver and passenger seats were adjustable, by sliding runners which no longer required wood packing strips. With the deletion of the longitudinal rail beneath the seats, the base of each seat no longer had to clear this obstacle.

The addition of door handles and locks may have spoiled the uncluttered lines, but provided a more civilised and secure car.

By this time, seats for North American cars used a modified frame in order to comply with local safety regulations.

All trim liner panels and full carpeting were colour-matched to the seats. The door liner panels, secured by spring clips, contained pale grey piping around the lower three-quarters of the cut-out area, the centre section of this being a separate steel panel covered in Vynide (matching the interior trim colour) and screwed directly to the door. A chromed window winder handle and chromed Mazak hinged door pull were provided. The chromed door release lever was sited at the top rear corner of the door in a recessed housing, which was finished in black with a chrome surround; the passenger side also had a locking lever. A door check strap was provided, as on earlier versions. The rear end of the door sealing rubber finished within a chromed capping piece at the top of the B-post.

Regardless of trim colour, the dashboard and all cockpit edging trim, including the door cappings, were finished in black. A crushable fibreboard parcel tray was now provided on the passenger side, and always finished in black.

The new hood gained a rather more vertical rear window than before. Swivelling quarterlight windows included the winding window guide channel.

A header rail, permanently fitted to the hood, secured the hood to the top of the windscreen frame. Two over-centre clips pulled the rail onto the frame with a rubber seal between. The strut between the top of the frame and the scuttle prevented the centre of the frame from being pulled upwards when the hood was clipped on.

ORIGINAL SPRITE & MIDGET

The tonneau now included a zip running inwards from the new B-posts, these now being required to allow the tonneau to be folded over the support rail in order to be fastened to the heel board studs.

The new hood frame had an additional link in front of the main bow.

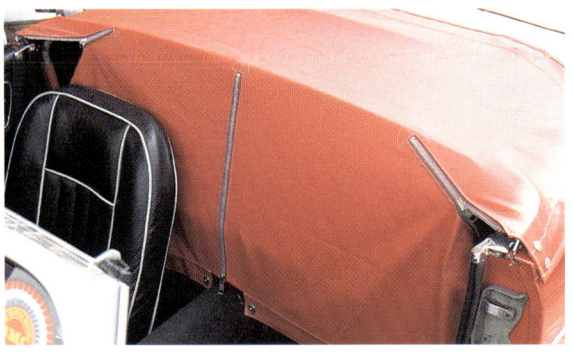

The revised tonneau cover (above), with zips running back from the B-posts, was still an optional extra. New door catch plates were required for the locking doors.

This is the genuine BMC optional hardtop (right) for the Sprite MkIII and Midget MkII. Alloy guttering directed rain water away from the door windows. Colours available were black, Riviera Blue, Tartan Red and Old English White.

DASHBOARD & INSTRUMENTS

The flat dashboard of previous models was now replaced by a shaped panel made up of two steel pressings welded together to form a single unit.

The speedometer and rev counter, still placed either side of the steering column, were now angled towards each other. The fuel gauge, previously on the outboard side of the panel, now moved to the shaped centre section of the new panel. All instrument and switch apertures were recessed into the panel.

The dashboard was always finished in black crackle paint, irrespective of body or trim colour. Three screws held the panel to the top scuttle rail with two support brackets at the lower edge. A small bracket was now included at either end of the panel to provide further support, each bracket being bolted to the top of the A-post beneath the scuttle. In the centre of the dashboard, either a 'SPRITE' or 'MIDGET' badge was applied, but the Sprite passenger no longer benefitted from a grab handle.

Instrument positions on a right-hand drive car were, from left: fuel gauge, combined oil pressure and water temperature gauge, rev counter (incorporating a red ignition warning light), speedometer with mileage and trip recorder (and incorporating a blue headlight main beam warning light). The positions for left-hand drive cars were, from left: speedometer, rev counter, fuel gauge, oil pressure/water temperature gauge.

All instruments were marked 'SMITHS', contained in chrome bezels, and calibrated in white on a black background. The rev counter had warning sectors in orange from 5500rpm and red from 6000rpm, was further marked 'POSITIVE EARTH', and carried the Smiths code number RV1 2401/00B. The speedometer carried a code number to indicate the calibration in relation to the rear axle ratio of 4.22:1, as follows: SN6142/00 (mph), SN6142/01 (kph), SN6142/02 (kph, West Germany, 145-13 tyres), SN6142/03 (kph, West Germany, 5.20-13 tyres). The speedometer trip reset was placed directly below the instrument, a small cutaway in the lower edge of the panel allowing easy access. The temperature gauge was marked in Fahrenheit or Centigrade, to suit local requirements.

The combined ignition/starter switch was placed

The tonneau support rail with the optional rear seat cushion and Kangol seat belt. The seat is the later type, as fitted from August 1965 onwards. The backrest could be altered for rake by adjusting two bolts at the base of the frame – not exactly reclining seats but useful nevertheless.

Sprite MkIII & Midget MkII

The second generation of seats fitted to the Sprite MkIII and Midget MkII after August 1965. Thicker backrests were provided and followed the style used in the MGB. The seat cushion was not removeable from the frame, so anything lost under the seats tended to stay lost!

centrally in the panel between the oil pressure/water temperature and fuel gauges. European versions with the combined steering column lock and ignition switch carried a chrome blank in the vacant dashboard hole from HAN8-40021 and GAN3-27195. A blocked oil filter warning light was placed above the oil pressure gauge, an unusual feature inherited from BMC's front-wheel drive vehicles with A-series engines.

Switches on right-hand drive cars were arranged as follows in relation to the ignition switch: to the right was a combined side and headlight toggle switch; to the left was the windscreen washer plunger; below the ignition switch and from left were the choke control (pull and twist to lock), windscreen wiper toggle switch (single-speed, self-parking), panel light toggle switch and heater control (push in for air flow, twist clockwise for blower). Only the heater and choke control functions were marked. For left-hand drive, the light switch and washer plunger were transposed, but otherwise the layout was the same.

Two green flashing indicator warning lights were fitted between the speedometer and rev counter. The indicators were now operated from a steering column stalk, to the right of the column on right-hand drive cars and to the left for left-hand drive. The indicators now self-cancelled with steering movement.

An option of a headlamp flasher became available, and was also controlled from the indicator stalk by pulling it towards the steering wheel. Both standard and optional stalks were chrome-plated with black plastic end. The dipswitch was, as before, floor-mounted to the left of the clutch pedal for both right- and left-hand drive. The horn was operated by the push button in the centre of the steering wheel boss.

Elbow room and door pockets were sacrificed to provide winding windows and lockable doors. The small locking lever was only fitted to the passenger door on the Sprite MkIII and Midget MkII.

The Sprite steering wheel with the Austin motif on the horn push.

79

ENGINE

BMC was very well aware of the design shortcomings of the 10CG engine in respect of crankshaft rigidity, so the problem was addressed in the 10CC engine by enlarging the main bearing diameter to 2in (50mm). Although the bore and stroke dimensions remained the same to give 1098cc, power output rose to 59bhp at 5750rpm with torque of 65lb ft at 3500rpm. This was achieved on a compression ratio of 8.9:1, although a lower ratio of 8.1:1 was available to suit local conditions, torque in this case reducing to 64lb ft at 3250rpm.

The lower oil control piston rings were now a Wellworthy-Duaflex 61 type. Pistons were available in overbore sizes of +.010in and +.020in. Although the cylinder block was no longer produced with an aperture for a camshaft-driven mechanical fuel pump, the front mounting stud for a mechanical fuel pump was still fitted and seemed to have no purpose.

To suit the closed-circuit crankcase breather system, the front tappet chest cover was now provided with an oil separator canister, a hose from this leading to the valve unit mounted on the inlet manifold. The rocker cover breaker pipe was now deleted, with a combined oil filler cap/breather fitted instead. Externally, this looked very similar to the previous black plastic cap, but contained a wire gauze breather element. From engine number 10CC-DA-H 25233, the rocker cover identification plates – MG, Austin and Weslake – were glued on rather than riveted.

In an attempt to cure the oil leakage problem from the tappet chest covers, there were new seals and gaskets along with a revised rear cover of improved stiffness from engine numbers 10CC-DA-H 37548 and 10CC-DA-L 25600.

The cylinder head contained inlet valves of 1.22in (31mm) diameter and exhaust valves of 1.0in (25.4mm) diameter. The porting was improved to increase gas flow, with the exhaust ports now served by a larger bore manifold and system.

Gold Seal replacement engines were available under the prefix 8G 150 for the 10CC-DA-H engine and 8G 149 for the lower 8.1:1 compression 10CC-DA-L engine.

COOLING SYSTEM

At an undefined point, the number of stiffening flutes in the radiator top tank was increased from three to four. Due to expansion and contraction, flexing had occurred in the top panel of the tank, sometimes causing cracking and leakage. The additional flute provided some improvement, if not a complete cure.

An optional factory-fitted oil cooler became available from around June 1965. The seven-row radiator was mounted to the bottom panel of the air intake, just in front of the water radiator. High-pressure hoses passed through the right-hand front panel and splash shield, which were now produced with corresponding holes above the heater air intake ducting; these holes would be fitted with blanking grommets when an oil cooler was not fitted. The pipes connected to the rear of the cylinder block and oil filter head, thus replacing the rigid oil transfer pipe.

Oil was pumped from the rear outlet, through the oil cooler and into the filter. There was no thermostatic control in the circuit, but a blanking plate was available to place in front of the cooler in cold conditions. When an oil cooler was specified, the water radiator fan cowling was fitted with a special bracket on the right-hand side to support the two pipes in the engine compartment.

CARBURETTORS & FUEL SYSTEM

A new cast alloy inlet manifold was introduced on the 10CC engine to allow the fitting of a Smiths closed-circuit crankcase breather system. A threaded tapping was produced in a boss on the top of the balance pipe between the inlet tracts. Into this screwed a steel stub pipe with a short section of ½in diameter hose connecting to the underside of the new breather control valve. This valve allowed crankcase fumes to be drawn into the inlet manifold and passed into the combustion chambers to be burned.

The condition of the rubber flap diaphragm within the control valve needed to be checked periodically. If allowed to deteriorate, this could upset the mixture strength from the carburettors by passing an excessive volume of air and crankcase fumes. Should it became perforated, then blue smoke from the exhaust and poor performance resulting from a weakened mixture could give the impression of a badly worn engine.

The twin HS2 SU carburettors were retained, but with a change of metering needles to types AN (normal), H6 (rich) and GG (weak).

Replacing the AC mechanical fuel pump was an SU electric pump, type AUF 216, with a maximum flow rate of 56pt per hour at a pressure of 2.5-

Dashboard and steering wheel were common to both the Sprite MkIII and Midget MkII, only the centre horn push and dashboard badge distinguishing the two. The gear lever cowl was now painted black regardless of body colour. A parcel shelf on the passenger side made up for the loss of stowage space in the doors.

A sticker applied to the speedometer glass informed drivers how to re-set the trip mileometer. No doubt most owners soon removed this sticker. Indicators were now operated by a column-mounted stalk. An option was a headlamp flashing facility operated by a visually identical but modified control, which was pulled towards the wheel to flash the lights.

SPRITE MKIII & MIDGET MKII

The much-improved 1098cc engine, type 10CC-DA-H or L. Power in high compression form was 59bhp at 5750rpm. The most obvious difference is the closed-circuit crankcase breather system with the Smiths diaphragm valve unit on top of the inlet manifold. The breather pipe between the rocker cover and front air filter was no longer required. Incorrect points to note are the transposed Austin and Weslake rocker cover plates, the heater box label and the battery earthing cable, which should have a black covering rather than being braided.

3.0lb/sq in. The pump was fitted adjacent to the right-hand rear shock absorber by a bracket attached to the heel panel and the underside of the axle tunnel. The pump was wired into the ignition circuit and would start and stop as the ignition key was operated. A small fuel filter was included on the inlet side of the pump. The pump carried a venting pipe which entered the boot space, and a T piece inserted in the end of this pipe prevented it from slipping through its grommeted hole.

Since the new crankcase ventilation system dispensed with the breather pipe between rocker cover and front air filter, the stub pipe connections were deleted from both.

EXHAUST SYSTEM

A new three-branch cast iron exhaust manifold with a longer downpipe mated to a larger 1.5in diameter (43.1mm) exhaust system. The connection between the manifold and the rest of the system remained as before, with a conical sealing face and split clamps.

The main exhaust pipe was re-aligned to move the single, straight-through, absorption-type silencer towards the fuel tank, the tailpipe now leaving the car to the right of the left-hand overrider. The round-bodied silencer was now constructed with crimped ends, whereas the small diameter exhaust system had flat end plates.

The torque reaction mounting between the bell-housing and exhaust pipe was also revised. A new bracket was attached to the two bottom bellhousing bolts, with a separate flat bar extending to the pipe and secured with a 'P' clip. A clip hung from the right-hand bracket bolt to secure the speedometer cable.

For Europe, a new twin silencer box system was introduced, the rear silencer being mounted transversely behind the fuel tank. New mounting brackets, supporting the pipe either side of the rear silencer, hung from the boot floor.

ELECTRICAL EQUIPMENT

The electrical system was carried over from previous versions with only minor changes, these mainly relating to the revised dashboard and control layout.

The cable-operated starter solenoid was now discarded in favour of an electrically-operated solenoid from the combined ignition and starter switch, although some European export cars already had this if a steering lock was fitted with a combined ignition/starter switch. The new solenoid was fitted in the same position on the heater and battery shelf, and contained a rubber-covered button which allowed the starter to be operated from beneath the bonnet.

Cars bound for the French market continued with a suppression screen around the distributor, while cars for Canada and Denmark were provided with rubber sealing boots to protect the coil, distributor cap and HT lead connection from moisture.

Headlamp specification continued to vary according to local regulations, but several changes occurred on cars for West Germany from HAN8-50192 (February 1965) and GAN3-38970 (March 1965). The sidelight bulbs were re-positioned into the headlamp units, leaving the front flasher units to be fitted with amber lenses of a concentric circle pattern radiating from the centre of the lens. The lenses were commonised to fit either side of the car. Also at this time, the red rear flasher lenses were changed to amber. Finally, two number plate lamp units, fitted to either end of a wider plate, replaced the single unit above the old plate. Bungs were applied to the redundant holes in the rear panel. North American cars continued to use red rear indicator lenses.

The sealed beam headlamp units remained unchanged, as follows (main beam and dip): RHD home and export, 60W/45W; LHD North America, 50W/40W. Bulb ratings for other markets were as follows: LHD except North America and Europe, 50W/40W; LHD Europe, 45W/40W.

TRANSMISSION

The gearbox carried over from the previous models, but from engine number 10CC-DA-H4642 and 10CC-DA-L2356 the gear tooth profile was changed from 'A' series to 'B' series. The two types are not interchangeable – an important point to remember when rebuilding a gearbox with replacement parts.

From chassis numbers HAN8-52666 and GAN3-40087 (June 1965), the propeller shaft grease nipples were deleted, the Hardy Spicer universal joints now being of the 'Sealed for Life' type. The gear lever cowl, formerly painted body colour, was now finished in black irrespective of body or trim colour.

REAR AXLE

The basic axle casing remained as before, but the spring, shock absorber and bump stop locations were revised to suit the new semi-elliptic springing arrangement. The upper radius arm was deleted as this was no longer required for longitudinal location.

The large bracket arrangement of the quarter-elliptic spring axle was discarded and replaced with a flat, horizontal platform below the axle casing to sit on top of, and to locate onto, the semi-elliptic spring. The shock absorber link no longer attached directly to the axle casing, the bracket provided for this being deleted. To the top of the casing, above the spring locating bracket and inboard of it, a new boxed bracket provided the location for a new cylindrical bump stop rubber buffer, the inner face of the box being provided with a stud to form the lower anchorage for the rebound check strap.

The axle casing for wire-wheeled cars was again 1in (25mm) narrower than that used for steel disc wheels. The differential gears and casing were unchanged with a ratio of 4.22:1.

FRONT SUSPENSION

The coil springs were changed to increase free length to 9.59in (243.6mm), taking effect from HAN8-58381 and GAN3-46042 (January 1966) for North America and HAN8-60260 and GAN3-47527 (March 1966) for all other markets. Other than this, the front suspension remained unchanged.

REAR SUSPENSION

Replacing the quarter-elliptic springs were semi-elliptic springs comprising five laminated leaves of 1½in (38.1mm) width. These were rated at 75lb/in with a free camber of 4.437in (112.7mm). The leaves were bound by six clamps with a round-headed bolt passing through the leaves in line with the axle location platform. The head of the bolt acted as a location dowel in the platform, which was suitably drilled with a mating hole.

The spring and axle were held together by two inverted ⅜ UNF threaded U bolts per spring, a clamping plate below the spring having four holes to accept the U bolts. In order to reduce noise and vibration, a rubber seating block was inserted between the axle platform and the top of the spring, and the clamping plate and the bottom of the spring. Each rubber block was contained by a pressed steel retaining piece, the U bolts passing through holes in these pieces to ensure their location. The lower clamp plate was produced with an upturned vertical flange face to the inboard side, the forward end of this containing a hole for the shock absorber link ball end to attach to.

The front end of the spring contained a Metalastic bush, a fulcrum bolt passing through this to secure the spring to the mounting plate. This plate contained four holes for the ⅜ UNF securing bolts; the two front holes had welded nuts for the bolts passing down through the floorpan immediately in front of the inner heel panel, while the two rear bolts engaged in threaded blocks welded to the inside of the rear bulkhead. The mounting plate bolted directly to the underside of the floor, accommodating the front of the spring within a recess in the rear bulkhead floor.

The rear of each spring was fitted with a conventional shackle arrangement with rubber bushing to upper and lower shackle pivots. The bushes, four per shackle, were inserted from each side into the spring eye and shackle mounting bracket, the bracket being secured to the boot floor by three 5/16 UNF Nyloc nuts and bolts.

The Lockheed disc brake, optional BMC anti-roll bar and splined wire wheel hub. Note the small hole in the splined portion to allow the castellated hub nut split pin to be fitted and removed – a tricky job even with long-nose pliers!

Wire wheels were a popular option, BMC offering these 60-spoke 4in × 13in wheels from the introduction of the HAN7 and GAN2 versions. At first these used a 12tpi thread for the hub nut, but this was changed to 8tpi for the HAN8 and GAN3. 'Eared' spinners, as shown here, were outlawed for many countries, home market cars being the last to change to hexagon nuts in 1969.

The Armstrong double-acting lever arm shock absorbers were mounted as before, to the inboard side of the gusset panel between the rear bulkhead heel panel and axle tunnel. No radius arm bracket was now fitted, so the shock absorber mounting bolts only passed through the gusset panel and shock absorber body. The shock absorber arm and drop link were suitably revised to relocate with the U bolt clamp plate.

As before, all suspension components were painted black, excepting the alloy shock absorber bodies which were left unpainted.

STEERING

The only changes for HAN8 and GAN3 models concerned the steering column and wheel.

A new column was provided with a revised upper portion to suit the new steering wheel and a self-cancelling indicator control stalk.

The new steering wheel, of 15½in (393.5mm) diameter, now contained six chrome-plated steel spring spokes arranged in pairs to give a formation of three main spokes. The wheel, finished in black, was now the same for both Sprite and Midget, although the central horn push motif was revised for both. This was now larger, the Midget using a correspondingly larger MG motif, the Sprite now having the Austin heraldic motif, an enlarged version of that used on the bonnet badge but without the lettering.

BRAKES

The braking system continued unchanged except for an alteration to the caliper securing bolts at HAN8-62388 and GAN3-50463 (July 1966). Plain hexagon-headed bolts replaced the earlier variety, which had been provided with a threaded extension to secure the flexible hose locking plate. The new bolts passed through the locking plates, which were now given correspondingly larger holes. A tab washer was now used to lock the head of each bolt.

Friction material was Ferodo DA3 for the front pads and Ferodo AM8 linings for the rear shoes.

WHEELS & TYRES

No changes occurred to either the standard steel wheel or optional wire wheel throughout this period other than a change of wire wheel hubs to include a coarser thread of 8TPI and the deletion of the wording '(NEAR)' and '(OFF)' from the spinner. Wire-wheeled cars bound for North America changed from 'eared' spinners to octagonal nuts during 1966 to comply with safety regulations, effectively leaving only right-hand drive cars with 'eared' spinners.

From August 1965, all Australian-assembled Sprites were equipped with wire wheels as standard, these cars being locally designated as Mk IIIA Sprites. Wheels were now painted Silver Birch Metallic in Australia, but 'Aluminium' continued elsewhere.

Dunlop radial ply tyres were now starting to become an option. Because of the slightly different rolling radius of these tyres, cars so fitted for the German market were equipped with recalibrated speedometers due to TUV regulations allowing only 5% speedometer error as opposed to the 10% of home market regulations. Dunlop C41 crossply tyres of 5.20-13 covered 929 revolutions per mile, while Dunlop SP radial tyres of 145-13 covered 934.

IDENTIFICATION, DATING & PRODUCTION

Jointly introduced in early March 1964, the Sprite MkIII and Midget MkII began with the prefix and chassis numbers of HAN8-38829 and GAN3-25788 respectively, the numbering sequence being a continuation from the HAN7 and GAN2 prefixed versions, thus demonstrating that Sprite sales were considerably higher than those for the Midget to this time.

Engines for the Sprite MkIII and Midget MkII were prefixed 10CC-DA-H or L, while Innocenti versions became 10CD. As before, a change of engine type returned the numbering sequence to 101, the last engine being 52908. Body numbers continued with an ABL prefix for the Sprite and GBE for the Midget, the body number plate being fitted to the left-hand A-post between the door hinges.

The Australian-assembled Sprite MkIII, which became available locally in October 1964, carried a new prefix of YHGN-8. The 'G' was the local code

Notes to table
From 1969, the Midget outsold the Sprite despite the latter being exclusive to the Australian market. Only North America favoured the Sprite when given a choice of either. There, the name of Donald Healey was probably held in higher regard than MG in association with these cars, no doubt due partly to competition successes.

PRODUCTION FIGURES

Year	RHD Home	RHD Export	LHD Export	LHD North America	LHD CKD	Total
Sprite MkIII (HAN8)						
1964	1773	147	1315	6562	608	10405
1965	2209	159	775	5095	644	8882
1966	1143	97	332	4902	144	6618
Total	5125	403	2422	16559	1396	**25905**
Midget MkII (GAN3)						
1964	3061	340	1695	5830	30	10956
1965	3626	278	844	4390	24	9162
1966	2597	131	516	3215	24	6483
Total	9284	749	3055	13435	78	**26601**

An original tool kit and bag with the optional grease gun, an item that no owner should be without if the steering and front suspension are to be properly maintained.

for an engine in the 1000 to 1399cc capacity range. Logically, the G should have replaced the A only when the 1098cc engine first appeared in the Sprite MkII HAN7, but there is evidence that these earlier Sprites carried both YHAN7 and YHGN7 prefixes. Home market coding allowed engines of up to 1200cc to be accommodated within the A code letter, but, as we shall see later, this was to be ignored. From August 1965, the Australian Sprite became the MkIIIA, with a new prefix of YHGN9; this marked the introduction of wire wheels as standard.

OPTIONS, EXTRAS & ACCESSORIES

The following items were available as factory-fitted options:

Heater
Fresh air unit
Tonneau cover, rail and stowage bag
Laminated windscreen
Whitewall tyres
Heavy-duty six-ply tyres
Dunlop SP radial tyres (145-13)
Wire wheels
Anti-roll bar
Headlamp flasher
Low compression ratio (8.1:1)
Rear compartment cushion
Hardtop (now common to both Sprite and Midget)
Oil cooler (available from June 1965)

These additional items were available as dealer-fitted options:

Wing mirrors (driver's side mirror standard for North America)
Luggage carrier
Seat belts
Auxiliary lamps
Locking fuel filler cap
Badge bar
Cigar lighter
Twin horns
Ace Mercury wheel discs
Chrome bumper guards (front and rear)
Supplementary tool kit (part number 97H 524)

Publication C-AKD 5097, still in loose-leaf form, continued to list the BMC Special Tuning options, including the resumed availability of engine tuning equipment. The 10CC-type 1098cc engine contained a far more robust crankshaft with 2in (50mm) main bearing diameter, which allowed the engine to sustain much higher power outputs than the standard 59bhp. Up to 74bhp at 6000rpm could be achieved using a 10.5:1 compression ratio and selected BMC Special Tuning parts. The following items (with part numbers) added to or replaced those listed for the MkII Sprite and MkI Midget:

Valve springs, inner and outer Total 140lb rating, AEA 768-7
Valve springs, inner and outer Total 165lb rating, C-AEA 494 & 524
Use with bottom cup AEA 403
Lengthening tappet-adjusting screw For use with standard rocker and high-lift cam, C-AEA 512
Lightened valve rocker AEG 425
Competition clutch cover and plate C-BHA 4448-9
Three-branch exhaust manifold C-AHA 5448
Strengthened rocker shaft C-AEG 399
Rocker spacer tubes to replace springs C-AEG 392
Lowered rear semi-elliptic road springs C-AHA 8272

Close-ratio gear sets were available. Care needed to be exercised to identify which gear tooth profile the gearbox was fitted with before modifying. At engine numbers 10CC-DA-H4642 and L2356 the gear teeth changed from 'A' series profile to 'B' series. Ratios were as follows:

	Helical-cut	Straight-cut
First	2.93:1	2.573:1
Second	1.754:1	1.722:1
Third	1.242:1	1.255:1
Fourth	1.0:1	1.0:1

PRODUCTION DATING

Date	Sprite	Midget	Notes
Jan 64	HAN8-38854[1]	GAN3-25825[1]	Production begins
Mar 64	HAN8-38829[1]	GAN3-25788[1]	Sprite and Midget jointly announced
Jan 65	HAN8-48873	GAN3-36791	First cars built in 1965
Jan 66	HAN8-57945	GAN3-45860	First cars built in 1966
Sep 66	HAN8-64734	GAN3-52389	Last chassis numbers
Oct 64	YHGN8	–	Sprite MkIII available in Australia
Aug 65	YHGN9	–	Sprite MkIIIA available in Australia
Nov 67	YHGN9	–	Discontinued

Note to table
[1] Although production began in January 1964, neither the first Sprite nor Midget by chassis number was the first built. The factory records give the following: HAN8-38854 and GAN3-25825 were the first cars manufactured, respectively on 28 January and 4 February; HAN8-38829 and GAN3-25788 were first chassis numbers, built respectively on 25 March and 23 March.

Sprite MkIII & Midget MkII

Found only on original BMC hoods are washing instructions, as shown here in the top left-hand corner of this Midget MkII's rear window.

PRODUCTION CHANGES

CHANGES BY CHASSIS NUMBER

HAN8-38829, GAN3-25788 (Jan 64)
First chassis numbers allocated.

HAN8-39635, GAN3-26642 (Mar 64)
Clutch hydraulic pipe shortened from 92in (233.7cm) to 71in (180.3cm) on LHD models.

HAN8-40021, GAN3-27195 (Mar 64)
Blocked oil filter light added to the dashboard above the oil pressure gauge.

HAN8-50192 (Feb 65), GAN3-38970 (Mar 65)
Changes for West Germany: front number plate bracket changed and two blanking plugs fitted to rear lower panel, front flasher lens changed to amber, sidelight relocated in headlight unit, rear light unit changed.

HAN8-52666, GAN3-40087 (Jun 65)
Propshaft grease nipples deleted.

HAN8-55373, GAN3-42995 (Aug 65)
Black seats changed, new frame and coverings.

HAN8-55518, GAN3-42979 (Aug 65)
Cardinal Red carpeting replaced with Red.

HAN8-55537, GAN3-43171 (Aug 65)
Blue seats changed, new frame and coverings.

HAN8-55625, GAN3-42827 (Aug 65)
Red seats changed, new frame and coverings.

HAN8-56903, GAN3-44165 (Dec 65)
Improved indicator self-cancelling spring catch.

HAN8-58381, GAN3-46042 (Jan 66)
Longer front road springs fitted, front number plate bracket changed on North American models.

HAN8-58701, GAN3-46562 (Feb 66)
Front number plate bracket changed on RHD models.

HAN8-60260, GAN3-47527 (Mar 66)
Longer front road springs fitted for all markets, in line with North American cars.

GAN3-49679 (Jun 66)
Radiator grille top and bottom rails and centre moulding improved, with revised MG centre motif.

HAN8-62388, GAN3-50463 (Jul 66)
Brake caliper mounting bolts and brake hose lock plates changed to delete lock plate nuts.

HAN8-62602, GAN3-50340 (Jun 66)
Boot lid badge spire clips replaced with plastic press-fit retainers.

HAN8-64734, GAN3-52389 (Sep 66)
Last chassis numbers allocated.

CHANGES BY ENGINE NUMBER
(all engines prefixed 10CC-DA)

101 (Jan 64)
First engine number.

H 1374, L 455 (Mar 64)
Rocker shaft pedestal plates changed, with cylinder head stud washers increased from five to nine.

H 3496 (Apr 64)
Improved core plugs fitted to cylinder block.

H 4642, L 2356 (Apr 64)
Gearbox gear tooth profile change from 'A' to 'B' series.

H 5279, L 2375 (May 64)
Dynamo adjusting link pillar and engine front plate changed.

H 6016, L 2391 (May 64)
Rocker shaft pedestal plates changed and head stud washers reduced to eight.

H 10800, L 2791 (Aug 64)
Improved seal to centre bolt for Purolator oil filter. Pressure switch included for by-pass valve warning light.

H 15980, L 13312 (Oct 64)
Reversion to original seal for above.

25233 (Mar 65)
MG, Austin and Weslake rocker cover plates now glued rather than riveted on. Revised oil filler cap.

27914
Improved oil thrower washer behind timing cover.

H 37548, L 25600 (Nov 65)
Gaskets changed for front and rear tappet chest covers. Improved rear cover introduced.

52908 (Sep 66)
Last 10CC engine.

COLOUR SCHEMES (HAN8, GAN3)

Body colour (paint code)	Seats/trim[1]	Seat piping	Carpet	Hood	Tonneau	Optional hardtop
Black (BK1)[2]	Black	Grey	Black	Black	Black	Black
	Red	Grey	Red	Black	Black	Black
Riviera Blue (BU44)	Blue	Grey	Blue	Blue	Blue	Riviera Blue
British Racing Green (GN29)	Black	Grey	Black	Black	Black	Black
Dove Grey (GR26)	Red	Grey	Red	Grey	Red	Black
Tartan Red (RD9)	Red	Grey	Red	Red	Red	Tartan Red
	Black	Grey	Black	Black	Black	Black
Old English White (WT3)	Black	Grey	Black	Black	Black	Old English White
	Red	Grey	Red	Grey	Red	Old English White
Fiesta Yellow (YL11)[3]	Black	Grey	Black	Black	Black	Black
Pale Primrose (YL12)[3]	Black	Grey	Black	Black	Black	Black

Notes to table
[1] Dashboard panel, front and rear cockpit trims and door top trims were always black irrespective of body or trim colour.
[2] Midget only.
[3] Sprite only: Pale Primrose replaced Fiesta Yellow in late 1965.

SPRITE MkIV & MIDGET MkIII
(HAN9/10, AAN10, GAN4/5)

Announced in October 1966, the Sprite MkIV (HAN9) and Midget MkIII (GAN4) featured an increase in engine capacity to 1275cc. Although the 1275cc engine had been introduced in April 1964 in the Mini Cooper S, the 65bhp version fitted to the 'Spridgets' differed in a number of important respects.

In order to improve reliability and reduce costs, early engines were redesigned to a new specification. Unfortunately, these early engines did not prove satisfactory, so the first 900 cars built were held at the factory to await improved units. Consequently, deliveries of 1275cc 'Spridgets' did not properly get underway until February 1967. This may also explain why some cars were final-assembled at Cowley between January and March 1967, in parallel with normal production at Abingdon, possibly as a means of catching up on orders. Although the engine did not become finally sorted until 1968, it gave a useful increase in performance.

A new folding hood and frame arrangement provided a much more convenient method of operation, the rear cockpit edge being moved 4in rearwards to accommodate the hood in the lowered position. The old combined brake and clutch master cylinder was replaced by separate cylinders in a new pedal box

Clive Wagstaff's Mineral Blue Sprite MkIV (HAN9) enjoys the view, as we do. Only the quarterlight mirror is non-original.

The front elevation of the Midget was virtually unchanged from introduction in June 1961 through to October 1969, only the sidelights being lowered slightly in the front wings in December 1968. Dave Gilbert's early GAN4 Midget has the 'high' sidelight position and non-original headlamp units.

Sprite MKIV & Midget MKIII

Only 21 'facelift' Sprites (HAN10) were finished with a black windscreen frame, so Stuart Granger's example must now be unique. It sports optional wire wheels and anti-roll bar.

The 'facelift' Sprite was built only for the home market, although Craig Polly's example has received North American style sun visors. The British Leyland front wing badge is missing, but then many enthusiasts felt it should never have been fitted in the first place!

assembly. And during 1968, the rear axle ratio was changed from 4.2:1 to 3.9:1 to take better advantage of the larger engine.

A significant development in November 1967 saw, for the first time, cars for the North American market having to be substantially modified to comply with new safety and emission control regulations coming into force from 1 January 1968. No longer could the controls simply be moved to the other side of the car to continue selling 'Spridgets' in their largest single market. Indeed, of the total number of Sprites and Midgets built between 1958 and 1979, some 76% were exported, showing the importance of complying with various local market regulations.

That same year, 1968, was significant for another reason: BMC and Leyland merged to form British Leyland. Triumph, makers of the Spitfire, and part of the Leyland group, now joined its great market rivals within the same company. Although this initially had little effect on the 'Spridgets' or the Spitfires, some startling changes came later on.

Facelifted versions of the Sprite and Midget were introduced in October 1969, their chassis number prefixes advancing to HAN10 and GAN5 respectively, although the model designations remained unchanged at MkIV and MkIII. These 'Spridgets' became truly badge-engineered versions of each other – gone were the small but identifiable details which separated the two. The Midget lost its vertical-slat radiator grille and side waistline trim strips, the bonnet strip having already disappeared in December 1968. The price of both versions became the same.

The HAN10/GAN5 facelift included revised interior trim, a new steering wheel, Rostyle road wheels and slimmer bumpers with rubber-faced overriders, the rear bumper being split to accommodate a square number plate on the rear panel. A twin silencer box

Original Sprite & Midget

exhaust system replaced the single box system, rear light lenses were flatter, and the number plate lamps were now mounted in the inner ends of the quarter bumpers.

Black trim featured in abundance on the exterior. The radiator grille was based on the simpler item used on the previous Sprites, but finished in satin black. The lower portions of the sills were black, and divided from the main bodywork colour by a trim strip. Early examples also received an epoxy-coated black windscreen frame, but this proved just too much for most customers and BL chose to return to the alloy-finished frame within a couple of months.

At the end of 1970, BL decided not to renew its contract with the Donald Healey Motor Co Ltd, thereby severing a long association between manufacturer and creator. Strangely, however, BL decided not to drop the Sprite at this time, continuing to offer this badged simply as an Austin Sprite with a new chassis number prefix of AAN10. The HAN10 and AAN10 Sprites were only built to home market specifications and sold in small numbers compared with the equivalent Midgets. Clearly, there was little point in offering the facelift version of the Sprite, so the final Sprite was assembled at Abingdon in July 1971.

The Midget was left to continue with minor changes until a more significant revision occurred in October 1971, when the rear wings were changed to give a full, rounded wheelarch with a small flared lip. The fuel tank was also deepened to increase capacity and new Rostyle road wheels were introduced. The home market version, not available until January 1972, acquired rocker switches instead of toggle switches on the dashboard. North American Midgets already had rocker switches as part of the safety package, along with side marker lights and three windscreen wipers. Still defined as a MkIII Midget with a GAN5 chassis number prefix, this version is commonly described as the 'round-arch' Midget.

BODY & CHASSIS

Immediately noticeable, the rear edge of the cockpit was extended 4in (102mm) rearwards to accommodate the new folding hood. The rear edge was now formed by a raised section with an upwardly turned flange to act as the securing face for the rear edge of the hood. The new hood frame was bolted to the inside of the B-posts through three captive nuts per side, the mounting sockets for the earlier hood frame being deleted.

The rear inner wings were now fitted with threaded bosses on the inside faces to provide seat belt anchorage points. In the engine compartment, the aperture for the master cylinder and pedal box was changed in order to accept the new arrangement of separate master cylinders for brake and clutch. From HAN9-70268 and GAN4-58112 (September 1967), the rear body panel was produced with apertures to accept twin reversing lights.

Up to now, bodyshells for either right- or left-hand drive had been the same, with duplicated apertures and welded nuts allowing assembly to either specification. Blanking plates and grommets were applied to cover the redundant apertures as required. However, from HAN9-72034 and GAN4-60441 (November 1967), cars for North America differed, to suit safety regulations, from those produced for the home market or supplied as CKD kits.

The North American bodyshell featured the following differences. A revised top footwell panel contained a mounting flange for the bottom of the collapsible, energy-absorbing steering column. The scuttle was revised to accept the new dashboard and facia panel. From HAN9-77591 and GAN4-66226 (December 1968), the bulkhead panel was modified to accept safety bonnet hinges. The scuttle panel now contained apertures for three windscreen wiper spindles, while each wing was drilled to accept front and rear side reflector 'Reflex' units.

For all markets, revised door assemblies were fitted from HAN9-72041 and GAN4-60460 (November 1967). These included a modified door latching mechanism, requiring the hole positions in the B-

As treasurer, at the time of writing, of the Midget & Sprite Club, Denis Matthews has to keep his head, hence the roll bar and later pattern head restraints perhaps? The 'round-arch' Midget is, arguably, the most popular and practical of all Sprites and Midgets, 'Frogeye' excepted. The Ochre trim tended to soil very quickly.

Early 1275cc cars had the heater air valve relocated to the front panel, a very long cable being required to operate this from the dashboard control. From HAN9-71121 and GAN4-58854 (September 1967), the valve was moved to the blower casing intake. The area behind the radiator grille should be finished black.

88

SPRITE MKIV & MIDGET MKIII

Very early 'facelift' cars (right) lacked a bracing gusset between the boot floor and rear panel bumper mountings. Above this area is the reversing light unit.

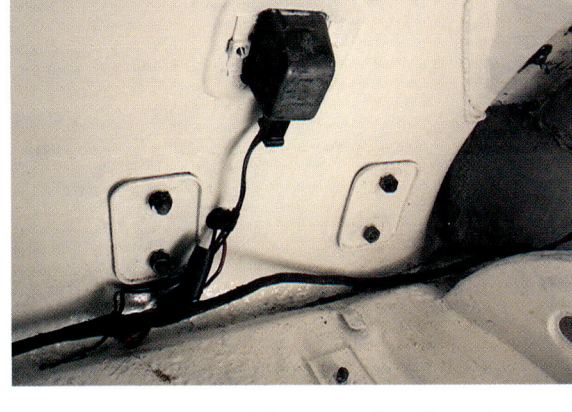

Two bracing gussets were soon added to either side of the rear panel (far right), similar to the single gusset per side of the cars with full-width rear bumpers.

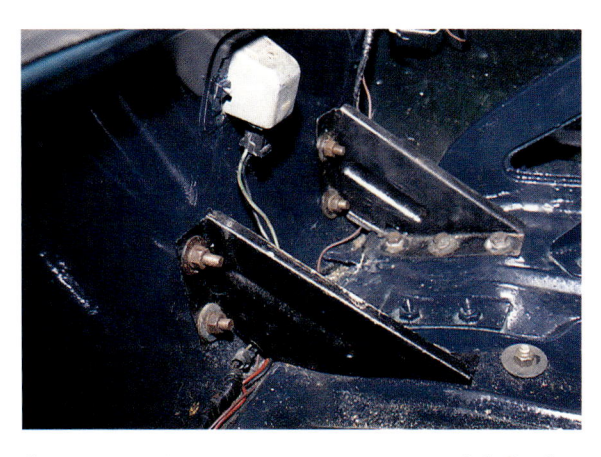

North American regulations specified that a certain area of the windscreen must be swept by the wipers. The answer for 'Spridgets' arrived with the addition of a third wiper blade from HAN9-77591 and GAN4-66226 (December 1968).

post to be altered slightly. Again for all markets, the front wings were produced with the sidelight apertures approximately 1in (25mm) lower down the panel from HAN9-77591 and GAN4-66226 (December 1968). Also at this time, the Midget lost its centre bonnet moulding strip, the bonnet no longer being drilled to accept the four retaining studs.

The facelifted versions of the Sprite MkIV and Midget MkIII, now prefixed HAN10 and GAN5 respectively, saw only minor revisions to the main structure. The rear wings were altered to accept new lamp units with flatter lenses. The rear panel was now produced with additional and repositioned holes for the new quarter bumper mountings, and the number plate light mounting plinth was deleted because new number plate light units were now mounted in each quarter bumper.

North American specification GAN5 bodyshells had an additional aperture in the scuttle panel for a third windscreen wiper spindle, while front and rear wings were modified to accept combined side marker light and reflector units. North American Midgets were also provided with additional holes in the boot floor to accept new venting pipes for the fuel tank, these connecting to a vapour separator tank bolted to the right-hand side of the boot.

At this stage, the Midget lost its waistline side trim mouldings, but both Sprite and Midget now gained a new moulding strip along the length of the sill. The sills also received further drillings to accept Sprite or Midget lettering. The bonnet was no longer produced with slots to allow the fitting of the Sprite badge, while both Sprite and Midget boot lid badging was changed with corresponding revisions to the hole positioning.

From HAN10-86301 and GAN5-89501 (June 1970), the bonnet and boot lid props were replaced with automatic-locking telescopic props which required new lugs to be welded to both panels, the right-hand front wing and the left-hand rear shroud. At HAN10-86303 and GAN5-89515 (July 1970), courtesy light switches were included for both doors and the boot lid, each requiring a welded bracket to be added to the body.

The heater unit was changed in September 1970 from a separate heater and blower to a combined unit, now secured to the bulkhead platform by six self-tapping screws replacing the four ¼ UNF welded nuts in the platform. The platform aperture was repositioned in the centre, rather than being offset to the left. This change, marked by body number rather than chassis number, occurred at GBE 100650 (home) and GUN 151300 (North America), separate body numbering having been introduced for home and North American bodyshells in November 1967. The body numbered changes tie approximately to chassis numbers HAN10-86377 and GAN5-91549 (home) and GAN5-91407 (North America).

The rear wings and inner wheelarch pressings were changed to give the round wheelarch shape from GAN5-105501 (October 1971). The export market took precedence on early deliveries, the 'round-arch' style not becoming available on the home market until January 1972. The inner wheelarch pressings were revised again on home market cars at GAN5-138801 (August 1973) to move the seat belt mounting slightly higher and forwards.

Anti-burst door locking took effect on North

ORIGINAL SPRITE & MIDGET

The square number plate, quarter bumpers, twin-box exhaust system and flat type light lenses are the clues that this is a 'facelift' Spridget.

American Midgets from GAN5-123731 (July 1972), calling for a revised door assembly and B-post to carry the locking post and plate. Doors for all markets were revised again from GAN5-138801 (August 1973), with mirrors standard on both doors.

Very late GAN5 floorpans contained an aperture below the rear of the gearbox. This appeared to be in readiness for the GAN6 bodyshell, which required access to the propshaft driving flange.

BODY TRIM

Both the HAN9 Sprite and the GAN4 Midget continued with the same external trim fittings as their predecessor models, but some detail changes were made during production.

Two Lucas reversing lights were incorporated in the rear panel either side of the number plate from HAN9-70268 and GAN4-58112 (September 1967). The centre bonnet moulding was removed from the Midget at GAN4-66225 (December 1968). This feature was now considered a safety hazard, following frontal impact testing during which, it is rumoured, a Morris 1100's similar moulding piece broke free and impaled the dummy driver. From HAN9-75703 and GAN4-64475 (July 1968), the quarterlight window assemblies were revised to incorporate re-shaped latch levers of square section.

This quarterlight revision took effect earlier on North American cars from HAN9-72034 and GAN4-60441 (November 1967), when sun visors and a driver's door mirror were added. The windscreen on North American cars became high-impact resistant laminated glass at the same time, although toughened glass was still standard for right-hand drive cars. North American and Swedish cars were fitted with reflector units to each wing, amber to the front and red to the rear, each unit being secured with two rivets into press-on fasteners.

The HAN10 and GAN5 versions, announced in October 1969, were the result of a cosmetic updating exercise which removed the essential differences between Sprite and Midget, other than badging.

The radiator grille was now common to both models, but based on that used on previous Sprites. Made of aluminium, it was finished in satin black with a single inset polished moulding piece. The badge for both versions was now mounted on a small central plinth in the grille and the Sprite bonnet was no longer fitted with a badge. The pressings surrounding the base and sides of the aluminium grille were now finished in a corresponding satin black, save for a narrow band of polished aluminium on the leading edge.

The front bumper, of slimmer section than before, carried a pair of smaller overriders with rubber-inlaid faces. The rear bumper, with matching overriders, was similarly reduced in section and split to give space for a square number plate to be fitted between the blades. The number plate was now lit by lamp units mounted within the inner end of each bumper blade.

The lower 4in (101mm) of the sills were now finished in black irrespective of body colour, with a dividing trim moulding piece between the black portion and main body colour. The black look initially extended to the windscreen frame with its black epoxy coating, but this feature met with such little enthusiasm that from HAN10-85307 and GAN5-77803 (October 1969) the frame returned to the familiar bright anodised finish. The wiper blades now contained a securing screw for the splined drive.

The boot lid badges were revised: the Sprite now received a round badge containing the usual heraldic Austin motif and the wording 'Austin-Healey Sprite', while the Midget was given an octagonal badge in black with polished MG lettering. British Leyland motifs were secured by spire fittings to the rear lower corner of each front wing, but from GAN5-118599 (April 1972) a single stuck-on motif was applied to the passenger side only and the colour changed from blue/silver to silver/blue. At the front of each sill Sprite or Midget lettering was applied: a single-piece badge served the Sprite, but the Midget had individual chromed letters retained by spires.

All GAN5 Midgets for the North American market had three windscreen wipers, and the side marker units, which previously contained only a reflector, now combined a light with a reflector.

From HAN10-86301 and GAN5-89501 (June 1970), soon after the facelift, an automatically-engag-

With the introduction of the 'facelift' Sprite HAN10 and Midget GAN5 in October 1969 came this black finish to the windscreen frame. The shortest-lived of all changes in 'Spridget' history, only 21 Sprites and 2918 Midgets were so finished before the bright frame was reinstated.

'Facelift' versions carried their front badges in the centre of the grille, which was now common to both Sprite and Midget but based on the earlier Sprite design. This example is an Austin Sprite (AAN10).

Like the Midget, the 'facelift' Sprite was treated to new badging.

SPRITE MKIV & MIDGET MKIII

New and much simplified badging accompanied the 'facelift' Midget (GAN5) from October 1969. The Midget script was deleted at this time: no doubt BL thought that the Midget had been around long enough to dispense with any further description for rear-end observers.

ing strut replaced the bonnet and boot lid props. The bonnet strut was pivoted from a lug welded to the right-hand side approximately halfway along the bonnet. As a result, the opened bonnet tended to distort and allow the left-hand front corner to dip, something that had previously occurred on the MkII Sprite and MkI Midget.

From 1 January 1971, the name Healey was deleted from the grille and boot badges of the Sprite, which continued in AAN10 form until production ceased in July.

No trim changes occurred when the Midget went into its 'round-arch' phase, but new quarterlight assemblies in stainless steel replaced the chrome-plated frames from GAN5-121650 (June 1972), and an anti-glare mirror was fitted as standard to each door from GAN5-138801 (August 1973).

In December 1973 from GAN5-143355, the over-riders on North American cars were replaced with large energy-absorbing rubber buffers as a first step towards the low-speed impact regulations coming into force. New front bumper mountings preceded this, being fitted from GAN5-138801 (August 1973). These drew extra support by extending rearwards to the suspension crossmember. The same front bumper mounting bracket was also introduced, on home market cars, but the small chromed overriders remained as before. At the rear, the larger rubber buffers dictated that new chromed licence plate lamp units be placed in a new position, either side of the licence plate on the rear body panel.

WEATHER EQUIPMENT

A completely new folding hood arrangement proved most effective and convenient to use, so much so that it went unchanged to the end of production.

Note the breather pipes (either side of the right-hand flute on the back of the axle tunnel) for venting the electric SU fuel pump body. Prior to HAN9-73294 and GAN4-61545, only one pipe was provided for this purpose.

The new hood frame, bolted to the inside of each B-post, formed a far more complicated system of links and pivots extending forward to join the header rail. The hood was permanently attached to the rear cockpit edge, being sandwiched between the mounting flange and a stainless steel retaining strip riveted through the hood and flange. Each of the seven rivets held a button fastener socket onto which the hood cover and optional tonneau cover could be attached. The previous padded roll edging was now deleted from the rear of the cockpit, so just the grey plastic – or black plastic from HAN9-77591 and GAN4-66226 in December 1968 – strip covering the flange was visible from inside the cockpit. Each end of this strip was covered with a chromed finishing cap over the cockpit flange, just above and behind the door sealing rubber finishing cap.

The rear quarters of the hood were secured to the body by four Tenax fasteners per side to the upwardly-turned cockpit flange. From HAN9-77591 and GAN4-66226 (December 1968), the front fastener on each side was replaced by a strip of Velcro to provide a more effective seal. When lowered, the hood and frame would fold into the rear of the cockpit, and a neat stowage cover could then be fitted.

The first taste of low speed impact absorbing bumpers arrived in December 1973. These substantial rubber buffer rear overriders (below) were for North American Midgets only, thank goodness. Front rubber buffers (right) and amber indicators feature on this 1974 Widget – sorry, Midget – in North American specification.

Original Sprite & Midget

Tony Slattery's YGN4 Midget demonstrates some variance between Australian and Abingdon-built Midgets. Abingdon would have deleted the central bonnet moulding at the same time as fitting the 'low-light' front wings. This Midget has the lowered sidelights which, to many people's surprise, pre-dated the introduction of the 'facelift' cars. Wire wheels, Dunlop radial tyres and a front anti-roll bar were all standard on Australian-assembled Midgets.

The longer cockpit of the 1275cc-engined versions produced a more rakish rear window than the rather upright style of the Sprite MkIII and Midget MkII. This Sprite, being later than HAN9-77591, has Velcro strips to retain the hood to the body just behind the doors. The quarterlight mirror has been added by the owner.

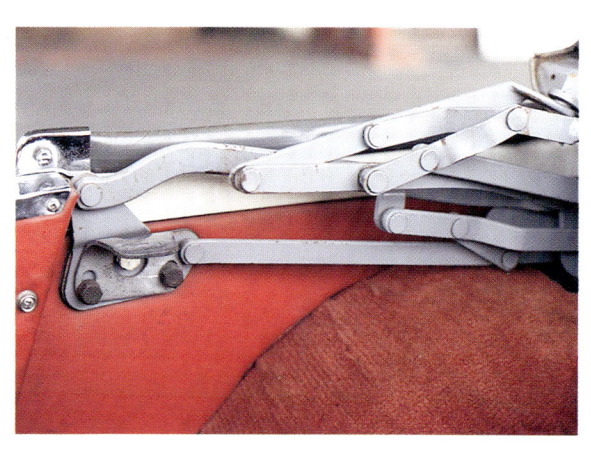

Providing that none of the pivots seized up, the new hood and frame arrangement (left) was very simple and quick to operate. The inward bend of the bracket provided a location for the revised tonneau rail. Note the grey plastic edging strip that ran around the cockpit edge to finish under a new chromed capping piece.

Earlier hoods (far left) had a Tenax fastener to secure the hood just behind the door.

92

Sprite MkIV & Midget MkIII

A hood cover appeared for the first time on the Sprite MkIV (HAN9) and Midget MkIII (GAN4) as the hood could not be removed from the frame or body for stowage. The cover buttoned to the rear cockpit edge and to the divider panel below.

The factory hardtop for the Sprite MkIV and Midget MkIII was based on that of the previous type, but with revised shaping to suit the new rear edging of the cockpit. Note that the lower edge of the hardtop is much straighter than before. This hardtop also continued to serve the Midget 1500 and was offered only in black.

The Velcro hood securing strip (right) that replaced the front Tenax fastener. Note the chrome capping piece and, to the rear, the tonneau support rail in place.

The new tonneau rail (far right) now engaged with a slot in the bracket which bolted to the hood frame mounting. After HAN9-77590 and GAN4-66225, rail, hood frame and plastic edging strip changed from grey to black.

From HAN9-77591 and GAN4-66226, the hood frame, header rail and rear cockpit plastic edging strip changed from grey to black. Note the webbing straps between the rear bows.

The cover used the same button and Tenax fasteners around the cockpit edge, while four straps on the inside contained further button fasteners to engage with corresponding sockets across the rear divider panel, totally enclosing the folded hood. To completely remove the hood assembly from the body, the rivets had to be drilled out and the frame mounting bolts removed from the B posts.

A new tonneau support rail fitted into L-shaped brackets bolted to the hood frame mountings, and, as before, the rail could be split for storage in its own bag. The rail, hood frame and header rail were initially painted Cumulus Grey, but from HAN9-77591 and GAN4-66226 (December 1968) these were painted black. The tonneau cover was correspondingly modified to suit the new method of attachment at the rear. The cover could later be obtained with pockets for head restraints, where these were fitted either as standard or as an optional extra according to the period and country of sale.

INTERIOR TRIM

Interior trim colour options at the start of HAN9/GAN4 production were reduced to black or red, according to body colour. The longer cockpit – necessary to accommodate the new folding hood arrangement – called for revised rear trim liner panels, the rear liner panel now sloping rearwards to meet the new position of the cockpit edge.

Before long, at HAN9-77590 and GAN4-66225 (December 1968), red seats and trim were discontinued, leaving black as the only interior colour. New seats, with horizontal fluting, were introduced at the same time. These seats could be reclined by operating a chromed lever placed on the inboard side of the squab. Cars for North American and West Germany were also at this time fitted as standard with head restraints, which had a twin-pole fitting into the seat backrest frame. Earlier on, door liner trim panels had been revised in November 1967, at HAN9-72034 (North America), HAN9-72041 (elsewhere), GAN4-60441 (North America) and GAN4-60460 (elsewhere). Black plastic window winder and door

Original Sprite & Midget

The first generation of reclining seat introduced from HAN9-77591 and GAN4-66226 (December 1968). A chromed handle on the side of the squab controlled the reclining mechanism, not visible in this view. Grey piping for seat and door trim was discontinued at the same time, along with red interior trim.

Revisions to the interior trim from HAN9-77591 and GAN4-66226 (December 1968) saw the white piping around the centre of the door liner aperture replaced with black. The window winder handle and door pull also changed from chromed Mazak to black plastic, and the door catches and striker plates were revised.

New interior trim for the 'facelift' (HAN10 and GAN5) appeared in October 1969. Heat-embossed pattern door liner panels were now in Black or Autumn Leaf, dependent upon body colour and date of manufacture.

The 'round-arch' Midget MkIII used a plastic strap door pull. Ochre trim replaced Autumn Leaf between July 1972 and August 1973, when Autumn Leaf returned. Not a popular colour perhaps?

pull handles replaced the previous chromed items, while the new liner panel contained black piping around the cut-out section.

The HAN10/GAN5 facelift versions announced in October 1969 featured a number of revisions to the interior and the deletion of the optional rear seat cushion. At first, black remained the only trim colour. The reclining seats were now covered with a heat-formed embossed pattern which contained a ladder effect in the centre section of the cushion and squab. These seats were also equipped with a fitting for single-pole head restraints, now an optional extra on home market cars. A blanking plug was fitted to the top of the backrest when no head restraint was fitted. This plug matched the seat trim when further colour options later became available, the first of these being Autumn Leaf in June 1970 from chassis number HAN10-86301 and GAN5-89501. The door pull handles remained in black plastic with the hinge mountings in chromed Mazak.

Facelift versions gained a black Vynide gear lever gaiter to replace the painted metal gear lever cowling and rubber gaiter. A semi-circular ashtray was fitted to Australian YGN4 Midgets at the front of the transmission tunnel, but an ashtray did not appear for other markets until the introduction of the facelift versions, when a tray with a hinged black lid was fitted to the transmission tunnel between the seats. Australian models, however, continued with the semi-circular ashtray, which was the same item as was fitted to similar period Land Rovers.

Door trim liner panels were changed again upon the introduction of the 'round-arch Midget at GAN5-105501. A full panel, without a cut-out, was of heat-formed Vynide, giving a simple horizontal ribbed pattern. The door pull handle was also changed to a flat strap, replacing the previous hinged handle. Black trim was replaced by a very dark Navy blue, but Autumn Leaf continued until it was replaced by the new colour of Yellow Ochre at GAN5-123731 (July 1972). Both of these colours were superseded at GAN5-138801 (August 1973), when Black and Autumn Leaf returned.

Although seat pattern and material remained the

94

The 'facelift' interior featured a new heat-embossed seat pattern, shown here with the optional head restraints. Note the ash tray on the transmission tunnel and the vinyl gaiter around the gear lever (although the bright retainer should be hidden beneath the carpet). The gear knob should be the same as on previous models.

same for all markets, the seat frame varied to suit local regulations or production improvements.

It became law to fit seat belts to all new cars in Britain from 1 January 1968, although mounting points had been included since 1961. The fitting of belts was initially left to the supplying dealer, but factory-fitted belts arrived in January 1971. Unless a customer specified otherwise, Kangol static belts were fitted. A storage bracket for the belt hook was provided on the hood frame body mounting bracket just behind the seat. Inertia reel belts were fitted from GAN5-116170 (March 1972) to North American cars, these carrying switches and wiring for a seat belt warning system. The reels and locking units were mounted to the rear inner wheelarch panels.

For the North American, German and Swedish markets, sun visors were fitted to the top screen frame rail from HAN9-72034 and GAN4-60441 (November 1967), and the rear-view mirror was mounted between them on the top screen rail, rather than on the tie rod which continued to be fitted. For other markets, the mirror continued to be mounted on the tie rod, but from HAN9-74462 and GAN4-63075 (June 1968) it was given a grey plastic frame to replace the previous black-painted steel frame. A dipping rear-view mirror was introduced for all markets on the facelifted HAN10 and GAN5 models. Set into a black plastic frame with the dipping lever at the bottom, this mirror was always mounted on the windscreen frame.

DASHBOARD & INSTRUMENTS

The basic dashboard and instrument layout continued unchanged for all right-hand drive HAN9/GAN4 models and early left-hand drive ones, although the instruments were subject to revisions.

A change from positive to negative earth on right-hand drive cars at HAN9-72041 and GAN4-60460 (November 1967) required a different rev counter, from Smiths RV1 2401/01 to RV1 2430/01. The fuel gauge was also changed at this point from FG 2530/70 to a half needle display type without any identification numbers on the face, the gauge now being controlled by a voltage stabiliser unit to dampen needle movement. For home market cars the temperature gauge now read simply H-N-C, replacing the Fahrenheit scale. Other markets continued with a Fahrenheit or Centigrade scale as appropriate until GAN5-105501 (October 1971), whereupon all cars were fitted with an H-N-C gauge.

Speedometer specification depended on the requirements of individual markets, and changed on all cars to suit the axle ratio reduction from 4.22:1 to 3.9:1 introduced at HAN9-77591 and GAN4-66226 (December 1968). Rev counters on all 1275cc versions contained an orange warning sector from 5500rpm and a red sector from 6300rpm. The various speedometers and rev counters used throughout HAN9/10 and GAN4/5 production are listed in the

Original Sprite & Midget

A short-lived feature of the first 'facelift' Sprites (HAN10) and Midgets (GAN5) was this style of steering wheel with angled upper spokes. The horn control also moved to the indicator stalk. From Sprite HAN10-86303 and Midget GAN5-89515 (July 1970), this wheel was replaced by a similar item with horizontal upper spokes, and the horn control on the centre boss. This very early 'facelift' Sprite features the redundant oil filter warning light lens above the oil pressure gauge – no bulb or wiring was fitted.

tables below and on the facing page.

A completely revised dashboard was installed for North American cars from HAN9-72034 and GAN4-60441 (November 1967). The facia and scuttle edge were foam-padded and covered in black Vynide with a vertical fluted pattern. The instruments and controls were arranged from left to right as follows: rotary control for the heater air valve; heater blower rocker switch (single-speed) with left-hand flasher warning light above; rev counter; combined oil pressure and water temperature gauge; steering column with brake circuit failure warning light above; fuel gauge; speedometer; side and headlight rocker switch with right-hand flasher warning light above; blue headlight main beam warning light with red ignition warning light and choke control above.

On the left of the steering column, a small toggle switch controlled the instrument lighting. From

From July 1970, the 'facelift' Sprite had a plain steering wheel boss which also contained the horn control. The padded rim steering wheel earned a reputation for imprisoning prying fingers in the holes of its spokes, which were now aligned horizontally.

GAN5-123751 (August 1972), this was replaced by a rheostat control positioned to the left of the choke and above the rev counter. Also at this time, all switches were given their own back-lighting for easy identification at night.

Two stalk controls were mounted to the steering column. To the left was a combined indicator, headlight dip and headlight flasher control. For a short spell, from GAN5-78446 to 89513, this stalk also controlled the horns, but either side of this period the horn control was in the centre of the steering wheel. On the right of the column, another stalk operated the two-speed wipers and electric screen wash. Also

SPEEDOMETERS

Smiths no.	Axle ratio	Application
SN6142/00	4.22:1	Mph
SN6142/01	4.22:1	Kph
SN6142/02	4.22:1	Kph, W Germany, radial tyres
SN6142/03	4.22:1	Kph, W Germany, crossply tyres
SN6142/04	4.22:1	Mph, W Germany, radial tyres
SN6142/05	4.22:1	Mph, W Germany, crossply tyres
SN6142/06	3.9:1	Mph, RHD from HAN9-77591/GAN4-66226
SN6142/07	3.9:1	Kph, LHD from HAN9-77591/GAN4-66226
SN6142/08	3.9:1	Kph, WGermany from HAN9-77591/GAN4-66226
SN5226/06	4.2:1	Mph, US/Canada
SN5226/09	3.9:1	Mph, US/Canada from HAN9-77591/GAN4-66226
SN5226/10A	3.9:1	Kph, Sweden
SN5230/02	3.9:1	Mph, LHD GAN5-105501 to 119495
SN5230/05	3.9:1	Mph, LHD GAN5-119496 onwards
SN6142/06 BS	3.9:1	Mph, RHD GAN5-141412 onwards

96

SPRITE MKIV & MIDGET MKIII

The door-operated courtesy light contained its own independent switch.

From GAN5-123731 (July 1972), this slotted-spoke steering wheel replaced the 'finger trap' design. Rocker switches appeared on home market 'round-arch' Midgets.

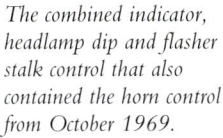

The combined indicator, headlamp dip and flasher stalk control that also contained the horn control from October 1969.

on this side of the column was the key-operated combined steering lock, ignition and starter control.

A centre console containing a central grille for a radio speaker accompanied the new North American facia. To the left of this, a rocker switch operated the hazard warning lights with a repeater light below. When the facelifted Midget arrived in September 1969, an optional console with a radio aperture could be ordered, and a cigar lighter was now fitted to the right of the speaker grille. This optional console became standard on North American 'round-arch' Midgets, and from GAN5-112276 (December 1971) a seat belt warning light was included above the cigar lighter. A radio and speaker console became an optional extra on home market cars after the introduction of the HAN10/GAN5 versions.

The bonnet release knob on North American cars was repositioned on the transmission bulkhead panel to the left of, and almost hidden by, the centre console, but right-hand drive cars continued with the release knob in a bracket to the left of the panel, just above the parcel shelf. A door-operated courtesy light with manual sliding switch was added to the lower edge of the facia, slightly covering the speaker aperture, from GAN5-89515 (July 1970). This modification also applied to home market facia panels without the centre console and included the Austin Sprite from AAN10-86803. On the passenger side, a cubby box with lockable lid completed the revised North American facia.

No significant changes were made to the dashboard of home market Midgets until the introduction of the 'round-arch' versions, when rocker switches appeared. The new switches – each with a white monogrammed function – were not illuminated, unlike the later type of North American switches. A minor rearrangement was necessary because these switches were larger, so the new positions, from left,

A combined ignition, starter and steering lock only became a standard fitting on home market cars in December 1970. The 'thin' type of head restraint seen here is incorrect for a GAN5 Midget.

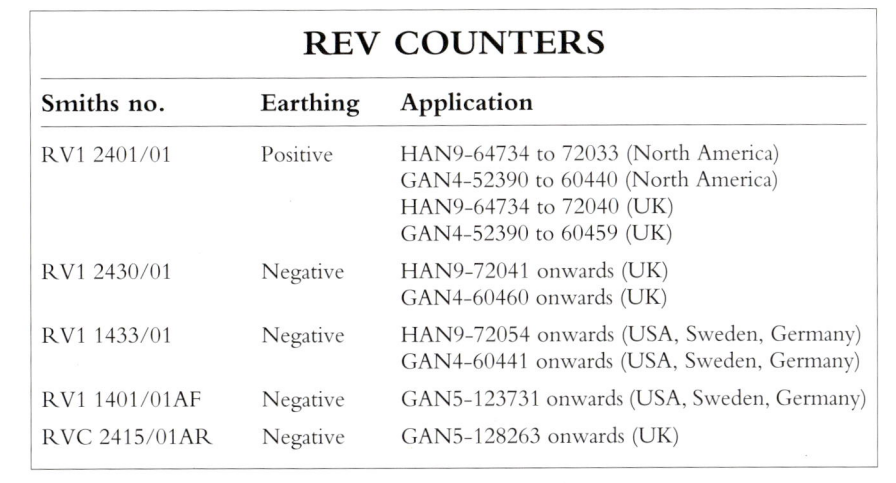

REV COUNTERS

Smiths no.	Earthing	Application
RV1 2401/01	Positive	HAN9-64734 to 72033 (North America)
		GAN4-52390 to 60440 (North America)
		HAN9-64734 to 72040 (UK)
		GAN4-52390 to 60459 (UK)
RV1 2430/01	Negative	HAN9-72041 onwards (UK)
		GAN4-60460 onwards (UK)
RV1 1433/01	Negative	HAN9-72054 onwards (USA, Sweden, Germany)
		GAN4-60441 onwards (USA, Sweden, Germany)
RV1 1401/01AF	Negative	GAN5-123731 onwards (USA, Sweden, Germany)
RVC 2415/01AR	Negative	GAN5-128263 onwards (UK)

ORIGINAL SPRITE & MIDGET

The first of the North American specification dashboards displays some fairly obvious differences from the home market cars. The gear lever cowl should be black.

were panel light switch, fuel gauge (as before), wiper switch (single-speed), oil and temperature gauge (as before) and side/headlamp switch. The manual windscreen washer pump was relocated to the left of the heater air valve/blower control, which remained as before.

All 'round-arch' Midgets used a combined ignition switch and steering lock, the dashboard no longer being produced with an aperture or blanking piece for the ignition-starter switch. Home market cars were not equipped with a hazard warning system until August 1973, at GAN5-138801, the rocker switch being placed to the right of the speedometer. The indicator repeater lights were revised to serve additionally as the hazard warning repeaters.

ENGINE

The introduction of the Mini Cooper S in 1963 demonstrated how the A-series engine could be made into a very capable performer with a higher state of tune than had been offered in any of the Sprites or Midgets. In the largest of its three forms, the Cooper S engine displaced 1275cc from a bore of 2.78in (70.61mm) and stroke of 3.2in (81.28m), producing a very healthy 75bhp at 5800 rpm on a compression ratio of 9.75:1.

The 1275cc Cooper S engine was announced in April 1964, just one month after the introduction of the MkIII Sprite and MkII Midget with the much-

The interior of this Australian-assembled Midget displays some of the non-Abingdon practices used down-under, such as the later type of door fittings with the earlier type of seats, still with grey piping. Note the standard Australian fitting of an ash tray (of a design common to Land Rover) on the transmission tunnel.

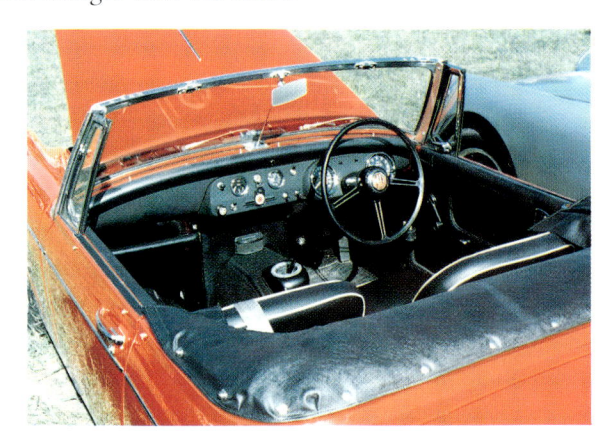

Another glimpse of the North American dashboard, this time in a 'facelift' version with the controversial 'finger trap' steering wheel (far left).

A 1974 North American dashboard with a totally idiot-proof steering wheel (left). Surely no-one could possibly get a finger stuck in this one?

The cylinder block on the right is an early 12CC type with a thin sump flange. Rigidity of the casting was somewhat suspect with the early engines. On the left, a 12V 778 FH block, the final type of 1275cc engine used in the last 12 months of 'round-arch' Midget production. Note the lack of oil pump priming plug and the fuel pump blanking plate.

The more commonly found EN16T tuftrided crank (left) and the highly prized EN40B nitrided crank fitted to only the very early 1275cc engines. Note the cross-drilling of the oilways on the EN40B crank to prevent oil being centrifuged to the outer bearing surface.

improved 10CC-prefixed 1098cc engine. With their attention immediately aroused, Spridget owners asked when they too could expect to see the 1275cc engine being offered. The answer did not arrive until October 1966 and the introduction of the MkIV Sprite and MkIII Midget.

At last, the specification gave the longed-for 1275cc but, sadly, not the 75bhp of the Cooper S. The high output of the Cooper S had made itself felt in terms of reliability with high oil consumption and a tendency for cracks to form in the cylinder head between the large inlet and exhaust valves. In order to produce a more reliable and cost-effective engine suitable for large-scale production, a significant redesign of the Cooper S unit resulted in a lower output of 65bhp at 6000rpm on a reduced 8.8:1 compression ratio.

The new 1275cc Sprite/Midget engine's cast iron cylinder block was instantly recognisable because it had no removable tappet chest covers, which not only improved the overall strength of the casting but also eliminated the traditional oil leakage problems associated with the two pressed steel covers used on all previous A-series engines. As with the 1098cc 10CC engine, no aperture for a mechanical fuel pump was produced in the casting of the 1275cc engines until the 1974 model year and the arrival of engines with a 12V 788FH prefix. Even so, these later engines did not carry a mechanical pump, the aperture being covered by a blanking plate.

The two central cylinders were 'siamesed', as the large bore size resulted in minimal wall thickness – without a water cooling space – between the bores. As in previous A-series engines, the crankshaft ran in three main bearings with end thrust washers fitted to the centre main. Very early engines, from 12CC-DA-H 1386 to 1397 and 1842 onwards, used a steel EN40B nitrided crank (more on this later), but this was replaced at an unspecified point by a Tuftrided EN16T crank. For 1974, the Tuftriding process was discontinued. The early EN40B cranks are highly prized for their performance, although the later cranks were by no means weak.

The front of the crank was now fitted with a twin (Duplex) timing gear and chain set to drive the camshaft, the cam sprocket no longer using two synthetic buffer rings to act as a chain tensioner. No separate tensioning system was used at all on these engines and, curiously, they seldom suffered the degree of chain rattle often heard on well-worn examples of earlier engines. The crankshaft pulley included a damper to absorb the vibration which the larger capacity imposed – the 1275cc engine was not one of the smoothest of the A-series family.

The camshaft, still running in three pressure-fed linered bearings, now had lobes increased in width to .500in (12.5mm). The oil pump, previously a peg drive from the rear of the camshaft, now became a flange type drive via an inserted adaptor fitted into the rear of the cam. The pump produced a higher running pressure of up to 70lb/sq in (4.9kg/sq cm) with increased flow rate.

Connecting rods were now split horizontally across the 1.625in (41.28mm) diameter big end bearings as the bores were now large enough to allow the rod to clear the cylinder walls when being fitted. The gudgeon pins were now a press fit in the connecting rod small end eye bushes, leaving the pistons to float about the gudgeon pin. Circlips were not required to retain the gudgeon pins within the piston. Dished crown pistons in aluminium alloy contained three compression rings and a Duaflex oil control ring. Pistons with increased dish provided a lower compression ratio of 8.0:1 as an option. As a result, torque dropped from 72lb ft at 3000rpm to 64.5lb ft at 3000rpm.

The cylinder head followed the familiar A-series configuration but was instantly recognisable as belonging to the largest capacity type by having a completely flat and machined upper face forward of the rocker cover. Previous heads had a cast finish with a slightly raised section below the thermostat housing. Head and block were not produced with the two additional holes and threads used to secure the head on the Cooper S. Valve sizes were 1.310in (33.27mm) inlet and 1.153in (29.29mm) exhaust.

Double valve springs, rated at 51lb outer and 25lb inner, were retained by split collets in a tapered seating cap; no spring clip was provided to further locate the collets above the cap. Valve stem oil seals were fitted to the inlet guides only.

It would appear that more than one casting pattern was used for the 1275cc cylinder head, although this is not identified in any of the parts books. The difference is in the shaping of the exhaust port around the valve guide; one head provides a step which conceals most of the pressed-in guide, whereas the other pattern is devoid of this step to allow much-improved port shaping with the guide exposed in the port. This second type is obviously more favourable for exhaust gas flow rate, but identification would appear to be a matter of physically checking this feature as all heads carried the casting number 12G 940 (the same as the late MkII and MkIII Mini Cooper S).

Ventilation was initially by closed circuit recirculator valve fitted to the inlet manifold, but, without the removable valve chest covers, the crankcase outlet was now located on the front of the timing cover casing.

The first generation of 1275cc engine carried the prefix 12CC, but this was replaced by the 12CE intermittently from HAN9-71937 and GAN4-60326 to HAN9-72670 and GAN4-61026 (December 1967), from which point all engines were of the 12CE type except those for North America. From November 1967, North American cars were equipped with emission controlled engines which were prefixed 12CD. Again, the engine numbering sequence began at 101 for both 12CE and 12CD.

The 12CE and 12CD engines were painted in a lighter shade of green with a hammer finish. This change is also thought to have marked the introduction of a stronger cylinder block casting, thereby bringing a final answer to the serious rigidity problems associated with the earlier 1275cc engines. It is possible that the nitrided EN40B crankshaft continued into the first few 12CE and 12CD engines, this crank being identified by the cross-drilling of the oil holes in the journals.

From 12CE-DA-H3201 and 12CD-DA-H8701, the closed circuit crankcase breather system dispensed with the control valve on the inlet manifold, a feature which, in service, could cause an excessive amount of

The 12CE-DA-H engine with vented carburettors to draw in crankcase fumes, this arrangement dispensing with the old Smiths control valve on the inlet manifold. Note the air filters with beveled rather than radiused corners.

The 12V 588FH engine of Denis Matthews' 1973 'round-arch' Midget displays the new air filter and rocker cover stickers. All 12V engines were painted black.

SPRITE MKIV & MIDGET MKIII

Believe it or not, this is the cleanest engine we have seen so far! Not externally perhaps, but this is a North American emission control engine. The air pump delivered air continuously to the exhaust ports and to the inlet manifold on the over-run. The rocker cover vent pipe suggests this is a 12CJ engine. A pipe would transfer fumes to a carbon absorption canister (missing here).

With North American evaporative loss regulations to be met, a fuel vapour separator tank was installed in the boot. This was placed in circuit between the fuel tank venting pipe and the carbon absorption canister in the engine compartment.

oil to be drawn into the manifold if the rubber diaphragm became cracked or split. Instead, a Y piece divided the breather pipe to vent directly into each carburettor body, revised carburettor bodies now incorporating a brass stub pipe for this purpose.

From August 1971, engines produced for the 'round-arch' Midget were painted black and carried a new prefix of 12V. This was followed by further identification as follows: 12V 586 FH, 101 to 6217, from August 1971 to December 1972; 12V 588 FH, 101 to 3240, from December 1972 to September 1973; 12V 778 FH, 101 to 2457, from September 1973 to October 1974. The black rocker cover carried a small British Leyland sticker on the right-hand side; the Weslake patents plate was not fitted to these engines.

The 12V 586 engine was unchanged from the previous 12CE. The 12V 588 received a new mounting bracket for an alternator, but was otherwise unchanged. The final engine, 12V 778, received a new exhaust manifold with a stud fitting for the exhaust downpipe. The cylinder block casting was revised to omit the oil pump priming plug, and to reproduce the mechanical fuel pump aperture and studs used on the 948cc and early 1098cc engines. The oil pump could be primed through the outlet tapping on the right-hand side of the engine just as effectively as using the priming plug.

Although provision for a mechanical fuel pump reappeared, a blanking cover was fitted as an electric pump was retained at the rear of the car. The reason for this feature was that basic casting and machining procedures were common to all 1275cc A-series engines being produced at that time for various British Leyland vehicles, some of which used a mechanical pump.

Gold Seal replacement engine prefixes were 8G 180 for all high compression engines (8.8:1) or 8G 179 for low compression 12CC or 12CE engines (8.0:1).

As a footnote, it is interesting to note that early production 12CC-prefixed 1275cc engines had a design fault which led BMC to replace the first 900 or so engines before the cars were sold. The fault concerned the rigidity of the cylinder block casting and crankshaft, prompting the temporary use of the expensive-to-manufacture EN40B nitrided steel crankshaft, backed up by a change of oil pump.

The real answer arrived in the form of an improved cylinder block casting which was less prone to flexing. This final improvement allowed the reversion to a more conventional, and cheaper, EN16T crankshaft, although this was given a Tuftriding heat-treatment process. Early cylinder blocks were subsequently known as thin-flange blocks, as later blocks had a noticeably thicker bottom flange to which the sump was bolted. From 1968, the internal wall thickness of the casting was increased around the bores, adding to overall rigidity.

The withdrawn engines were between numbers 12CC-DA-H101 and 1000, and these numbers were not re-issued. Very early cars, therefore, have higher engine numbers than would be expected.

Because of the engine problem, very few, if any, Sprites or Midgets reached private owners between announcement in October 1966 and the end of the year, when the modified engines were ready. Indeed, February 1967 appears to be the earliest that deliveries got properly under way, a factor which may also account for the duplication of assembly at Cowley as a means of catching up on orders. This, I should stress, is my personal theory; there may have been other reasons why Cowley was employed to assemble Sprites and Midgets between January and March 1967.

NORTH AMERICAN ENGINES

From 1 January 1968, cars sold in North America were required to meet new emission control regulations. This necessitated revisions to the 1275cc engine in order to reduce hydrocarbon and carbon monoxide levels by injecting air directly into the exhaust ports. A modified cylinder head casting was provided with drillings from the right-hand side into each exhaust port. Each raised boss between the spark plugs contained an injector tube and threaded hole for the air delivery pipe connection.

Air was supplied by a pump driven by belt from a second groove on a new water pump/fan pulley. The air pump was pivoted from a revised water outlet casting with a flat link adjustment bar beneath the pump to tension the driving belt. Air was drawn through a replaceable paper element filter into the vane type pump, the compressed air then being passed through a non-return check valve behind the pump. The check valve would prevent exhaust gas reaching the pump should air pressure fall below that of the exhaust. From the check valve, a pipe would then deliver air to each of the four cylinder head injectors. The pump also contained a blow-off valve to limit maximum air pressure.

Between the pump and check valve, a T piece allowed air to be passed to the induction side of the engine. A gulp valve placed towards the front of the engine was controlled by a vacuum pipe from the inlet manifold. When the throttle was closed and a high manifold depression created, the gulp valve would open, allowing air to be drawn into the inlet manifold and thus weakening the mixture on the over-run. Air entered the manifold through a T piece below the crankcase ventilation recirculator valve in the centre of the manifold balance pipe.

This emission-controlled engine was identified by the prefix 12CD-DA-H (101 onwards), the compression ratio remaining at 8.8:1. Power output was reduced by 3bhp, this being the amount required to drive the air pump.

At engine number 12CD-DA-H 8701, the recirculatory crankcase ventilation system was revised to delete the inlet manifold vacuum control valve and to vent the crankcase directly into each carburettor. 'Ventilated' carburettors were now fitted, similar to those on the domestic 12CE engine, with a Y piece connecting each carburettor to the crankcase vent pipe from the timing cover. The air delivery pipe from the gulp valve now entered the inlet manifold via an elbow.

An addition to the emission control regulations concerned evaporative loss control – the release of fuel vapour into the atmosphere. Introduced at the same time as the Midget's GAN5 facelift, in October 1969, the evaporative loss engine received the prefix 12CJ-DA-H and began at number 21201. Initially, this engine was possibly only fitted to cars destined for California, while the 12CD continued for all other Federal states. However, the 12CJ would have been adopted for all North American states for the 1971 model year, commencing at GAN5-89596 (August 1970).

A new fuel tank was fitted with a non-venting filler cap and, within the tank, a capacity-limiting chamber prevented fuel expansion from pressurising the tank. A breather pipe from the tank led to a vapour separator in the boot, a further pipe from this leading to a carbon absorption canister mounted in the engine compartment. Also connected to the canister was a pipe venting both carburettor float chambers. The canister was purged by an interconnecting pipe to the rocker cover, any vapour from the fuel tank and float chambers being stored in the canister and then drawn into the engine through the purge pipe.

In August 1971, for the 'round-arch' Midget, the compression ratio was reduced to 8.0:1 and an alternator replaced the dynamo. Power output fell to 54.5bhp at 5500rpm and torque became 67lb ft at 3250rpm. A new prefix of 12V 587 ZL (numbered 101 to 15393) applied until August 1972, whereupon the notation 12V 671 ZL (numbered 101 to 23958) took over until October 1974.

Gold Seal replacement engine prefixes for North American 1275cc engines were as follows: 8G 189 for high compression 12CD and 12CJ; 8G 190 for low compression 12CD and 12CJ; 38G 506 for 12V 587 ZL and 12V 671 ZL.

COOLING SYSTEM

The 1275cc engine utilised the same cooling system as before, but now with a six-blade steel fan. This was in turn replaced by a six-blade plastic fan (still in yellow) from engines 12CE-DA-H899 and 12CD-DA-H101 (North America). This coincided with the change from a vertical to a cross-flow radiator, which was necessary to allow space for fitting an air injection pump on the right-hand side of the engine for the North American market.

The smaller cross-flow radiator was fitted into a black-painted sheet steel cowling. The inlet and outlet stubs were now positioned on the opposite sides, requiring the bottom hose from the right-hand side of the radiator to connect with a transverse cross-over pipe bracketed to the rack and pinion mountings. This steel cross-over pipe contained a drain plug screw as the new radiator was not provided with any means of draining. On the left-hand side, a second bottom hose connected the cross-over pipe with the

The separate heater and blower units were replaced during September 1970 by a combined unit. This required a repositioned aperture in the shelf below to allow the airflow into the plenum chamber for distribution to either footwell or demister slots in the scuttle.

With the optional oil cooler, the hoses pass through the splash shield and front air intake panel to the cooler, mounted in front of the radiator. The hoses were secured by this bracket to the inner wing for crossflow radiator cars, or by a bracket to the fan shrouding panels with the vertical flow radiator.

water pump. The temperature gauge transmitter bulb was now fitted into a tapping in the right-hand side of the cylinder head just below the thermostat.

A new thermostat housing, now pointing to the left, was equipped with a longer and reshaped top hose to connect with the radiator inlet stub on the left-hand side, North American cars having a modified housing to include a mounting boss for the air pump. The cross-flow radiator was connected to a small black-painted expansion tank mounted on the left-hand splash shield. A small-bore hose connected the radiator to the tank, which was fitted with a 15lb sq/in pressure cap.

The radiator was filled through a hexagon-headed plug screwed into a threaded boss set in the right-hand corner. The head of the plug also carried a wide screwdriver slot. A warning sticker was applied to the cowling stating that the filler plug should not be removed while the engine was hot. Water capacity was reduced to 6 pints (3.4 litres).

The precise change-over point from vertical to cross-flow radiator is not stated in the factory parts book, but the re-positioning of the windscreen washer bottle, necessary to accommodate the positioning of the expansion tank, occurred at HAN9-72034 and GAN4-60441 (November 1967), as did the arrival on cars for North America of air injection, which had to be accompanied by the cross-flow radiator. It is possible that some home market cars, or cars for other export markets, still used the vertical-flow radiator for a short time after this.

At around the same time, the water pump was revised to include a larger diameter inlet stub with an accompanying change to the bottom hose. The lubrication plug in the right-hand side of the pump casting was no longer provided as the bearings were now sealed for life. This change took effect from engine number 12CE-DA-H874, slightly preceding the change to the cross-flow radiator on UK cars.

Wax type thermostats of 74°C (165°F), 82°C (180°F) or 88°C (190°F) were specified according to climatic conditions.

Some changes occurred to the heater system, which was still an optional extra. Up to HAN9-71120 and GAN4-58854 (September 1967), the air control flap valve had been positioned in the front panel assembly and operated by a long Bowden cable from the dashboard control. This long cable control did not prove very satisfactory in service, so a new inlet tube containing the flap valve was mounted to the front of the blower. The heater and blower units were revised at HAN9-72034 and GAN4-60441 (November 1967), North American and Swedish market cars having a new rotating dashboard control for the flap valve and a separate rocker switch for the single-speed fan blower; UK cars continued with an unchanged single control. The fresh air unit remained an option for some hot countries.

The crossflow radiator and separate expansion tank reduced water capacity but operated at a higher pressure. The change from vertical to crossflow was necessary to allow North American emission control equipment to be fitted. The fan was also changed from steel to plastic at this point.

A heater became standard on the facelifted HAN10 and GAN5 versions, except on the Australian market where the fresh air unit became standard on YGN5 Midgets. Heater units for Australia were after-market items: the factory heater system, which could be ordered through dealers, is rare to find because it was expensive to order and fit, but some cars received locally-sourced units.

The heater and blower units were changed once again in September 1970 at body (not chassis) numbers GBE-100650 (home market) and GUN-151300 (North America). These points tie approximately to chassis numbers HAN10-86377 and GAN5-91549 (home market) and GAN5-91407 (North America). The new heater assembly contained the blower fan and water matrix within a single black-painted steel casing. The casing, still held together with spring clips, was now secured to the platform by six self-tapping screws. Inlet and outlet stubs for the matrix were now both to the front left-hand side of the casing. The cylinder head heater tap was turned through 180° to provide an easier connection, via a rubber hose, to the relocated inlet stub. The larger heater/blower casing required the starter solenoid to be moved to the right-hand upper footwell panel.

This new heater provided a neater installation and a stronger output of warm air. The cold air ducting pipe from the front of the engine compartment was slightly re-routed at the rear to meet up with an angled air inlet stub which also housed the air flap valve. The blower fan was still single-speed and the cockpit controls were unchanged. The new heater displayed a sticker to state that the unit was impossible to drain when the remainder of the system was drained, and that Bluecol anti-freeze must be used in freezing conditions.

CARBURETTORS & FUEL SYSTEM

The fuel system was carried over from the previous versions except for the air filter casings now having beveled rather than radiused outer corners to improve clearance with the bonnet, the 1275cc engine being slightly taller than the 1098cc engine.

Shortly after the introduction of the 12CE engine, the crankcase breathing system was changed. The venting pipe now fed directly into each carburettor, thus deleting the diaphragm-controlled venting valve on the inlet manifold balance pipe. The carburettors each carried a brass tube in the main body on the manifold side of the butterfly disc valve. The tubes, angled upwards between the carburettors, were joined by short lengths of hose to a Y piece, which was then connected via a further hose to the breather canister on the timing cover.

On North American specification engines with an air injection pump, air from the gulp valve entered the inlet manifold, initially via a T piece below the close-circuit crankcase venting valve. In accordance with the deletion of this latter valve on engines without emission control, the 12CD-DA-H engine was similarly affected at engine number 8701, the air pipe from the gulp valve now connecting to an elbow adaptor on the inlet manifold. A vacuum-sensing pipe, connected to a tapping in the inlet manifold, controlled the opening and closing of the gulp valve. High manifold depression, created on an overrun throttle, opened the valve and allowed the air pump to deliver air into the inlet manifold, thus weakening the mixture and reducing pollutants when slowing from speed. The air pump also delivered air into each exhaust port, but this function was not controlled by inlet manifold depression. At engine number 12CE-DA-H3401 (plus 3201 to 3300), a revised inlet manifold was fitted.

As explained earlier, North American evaporative loss regulations were met by the 12CJ engine. On this, the float chambers were fitted with vent pipes which were connected to a carbon absorption canister in the engine compartment. This was located on the passenger side upper footwell panel, the right-hand drive master cylinder pedal box blanking plate now having a bracket to carry the canister. The canister contained carbon granules to absorb vapours and an air filter pad in the base. The filter required renewal every 12,000 miles and the complete canister had to be changed every 24,000 miles as the carbon became saturated.

The fuel tank was fitted with a non-venting chromed filler cap and an internal capacity-limiting chamber. A vent pipe from the tank connected to a vapour-separating tank bolted to the right-hand side member in the boot. From here, a further vent pipe led to the engine compartment to connect with the absorption canister. The canister was purged by a pipe from the rocker cover to draw collected vapours into the engine.

The vented SU carburettors for the North American 12CD and 12CJ engines were fitted with swing needles rather than the fixed ones used hitherto on all other carburettors. The swing needle, type AAC, was less sensitive to wear if the jet and needle were not centred properly — always a tricky operation to perfect with the fixed needle jet. The later 12V engine for the North American market, types 12V 587 ZL and 12V 671 ZL, both with the 8.0:1 compression ratio, moved on to type AAT or ABC swing needles. These Midgets now carried an emission control information sticker on the right-hand side of the bonnet slam panel. Swing needle carburettors did not appear on the home market until September 1973 and the introduction of the 12V 778 FH engine.

Emission control engines from 12CD-DA-H8701 onwards all used carburettors fitted with throttle disc valves to limit manifold vacuum when closing the throttle from high revs. These engines were also served by an SU AUF 305 fuel pump which complied with evaporative loss regulations.

The accompanying table lists the carburettor variations used on all of the 1275cc engines, along with the number shown on the identity tag screwed to the float chamber lid.

The black-painted air filter casings remained the same on all 1275cc engines, although the yellow Cooper script gave way to a Unipart Filters sticker with the introduction of the 'round-arch' Midget.

On earlier cars, the heat shield had been prone to cracking below the manifold, due to the action of the three throttle return springs tending to bend the shield as the throttle was opened and closed. In time, the shield could be weakened sufficiently to bend, or even break, causing the throttle return springs to lose tension completely. This problem was addressed with the introduction of the 12CD and 12CE engines, with the addition of bracing brackets to the lower corners of the shield, the front bracket connecting to the engine mounting/timing cover plate and the rear bracket to a bellhousing bolt. With the introduction

Notes to table
[1] Non-venting to 3400 excepting 3201 to 3300, as per 12CC-DAH carburettors.
[2] North American emission control engines.

CARBURETTOR IDENTIFICATION

Engine prefix	Carburettor type	Needle	Front carb	Rear carb
12CC-DAH	Non-venting	AN	AUC9268	AUC9269
12CE-DAH[1]	Venting	AN	AUC9029	AUC9030
12V 586 FH	Venting	AN	AUC9029	AUC9030
12V 588 FH	Venting	AN	AUC9029	AUC9030
12V 778 FH	Venting	AAC	CUD9296	CUD9297
12CD-DAH[2]	Non-venting	AN	AUC9460	AUC9461
12CD-DAH (from 8701)[2]	Venting	AAC	CUD9031	CUD9032
12CJ-DAH[2]	Venting	AAC	CUD9081	CUD9082
12V 587 ZL[2]	Venting	AAT or ABC	CUD9180	CUD9181
12V 671 ZL[2]	Venting	AAT or ABC	CUD9180	CUD9181

Sprite MkIV & Midget MkIII

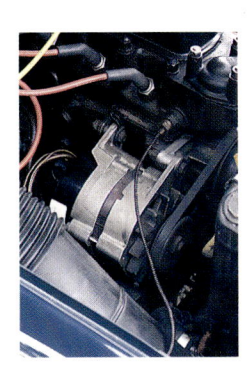

The 12V 588FH engine with the Lucas 17ACR alternator. The high tension leads would originally have been black.

of the emission control engines, 12CD onwards, a small section had to be removed from the front top corner of the shield to provide clearance for the gulp valve and pipework – all shields were thereafter produced with the cutaway.

The fuel tank sender unit was changed to a flat top type when the earthing polarity was changed to negative at HAN9-72041 and GAN4-60460 (February 1968), the gauge now being in circuit with a voltage stabiliser. From GAN5-105501 (October 1971), tank capacity was increased to 7 gallons (32 litres) and the sender unit changed again, now being retained by a locking ring instead of the previous six-screw fitting. These larger tanks were 1in (25mm) deeper than the earlier 6 gallon (27.3 litre) tanks. Evaporative loss tanks contained approximately half a gallon (2.3 litres) less because of the capacity-limiting chamber.

Exhaust System

No changes to the exhaust system occurred until the HAN10 and GAN5 versions were introduced in October 1969.

Except for North America, a new twin-box tail system (similar to that used on European HAN8 and GAN3 cars) replaced the single silencer at this point, but retained the same front pipe and manifold. A smaller resonator box occupied the position of the former silencer, the new silencer being placed transversely behind the fuel tank with the tail pipe turning through a right angle to leave the car on the right-hand side. The new tail system was a complete assembly with no joints other than a sleeve fit to the front pipe. Two new mountings below the boot floor mated up to the inlet and outlet pipe of the main silencer, but from GAN5-138801 (August 1973), both right- and left-hand rear mounting brackets for the transverse silencer were revised.

A new exhaust manifold was introduced for the emission-controlled North American cars with the 12CD-prefixed engine from HAN9-72034 and GAN4-60441 (November 1967). This was now produced with a flanged face-to-downpipe joint, with a corresponding change to the downpipe, which now had a flared sleeve to locate a sealing ring. A loose-fitting flange plate around the downpipe was drawn upwards by three nuts and bolts through both flanges to compress the sealing ring. This provided a much more positive and secure means of mating the downpipe to the manifold – a long-time problem with the conical fitting which continued on the home market cars.

From HAN9-72789 and GAN4-60694, North American cars were fitted with a single-piece exhaust system. Home market cars did not receive the flange fitting manifold and downpipe until GAN5-139137 (August 1973), but now with studs to replace the three bolts. From GAN5-139773 (September 1973) both home and North American Midgets now used a single-piece, twin-box exhaust system.

Electrical Equipment & Lamps

Initially, the battery remained a Lucas N9 (or dry-charged NZ9 for export) of 40 amp/hr capacity. The Lucas C40 dynamo, RB106 control box, 4FJ fusebox and M35G starter were carried over from the previous versions. The windscreen wiper motor, for all markets, was repositioned further outboard, to clear the new pedal box arrangement when equipped for left-hand drive.

The distributor for engines prior to 12CC-DA-H 11638 was type 23D4, which did not have a vacuum advance unit or micro-adjuster. After this engine number, the type 25D4, as fitted to the earlier 10CC engine, was reinstated. Spark plugs were now Champion N9Y with a Lucas 11C12 coil.

Twin Lucas reversing light units were incorporated in the rear body panel from HAN9-70268 and GAN4-58112 (September 1967) and were automatically operated from a switch built into the gearbox remote control housing. From HAN9-71624 and GAN4-59608 (October 1967), the starter motor solenoid was changed, the replacement not including a manual starting button.

A change from positive to negative earthing occurred at HAN9-72034 and GAN4-60441 on North American models and HAN9-72041 and GAN4-60460 elsewhere. The battery was changed to a Lucas A9 (40 amp/hr) or alternative A11 (50 amp/hr), equivalent dry-state batteries for export being AZ9 or AZ11; the new batteries had translucent bodies to make it easier to check the electrolyte level. A revised spill tray (with very low, straight sides), clamping bar and rods accompanied the new battery, which had revised main cables although the terminals remained at the rear. A negative earth warning sticker was applied to the bonnet slam panel.

The change of polarity required different rev counters according to market (see the 'Dashboard & Instruments' section), the new instruments being marked negative earth. The fuel gauge was also wired through a bi-metal voltage stabiliser to prevent rapid oscillations of the needle. The Lucas wiper motor changed from a square-bodied DR3A to a round-bodied 14W, wired on North American models for two-speed operation with the column stalk also controlling an electric washer pump mounted on the bulkhead. The ignition coil changed to a Lucas HA12, the low-tension terminals having '+' and '–' markings to replace 'CB' (Contact Breaker) and 'SW' (Secondary Winding) markings on the previous positive earth coil.

From HAN9-77591 and GAN4-66226 (December 1968) on North American cars, the brake circuit failure switch and dashboard warning light (with bulb-testing switch) were revised, each having an additional connection to improve reliability. The Lucas starter motor changed from M35G to M35J at an undefined time during the run of the 12CE engine between January 1968 and August 1971. Another unspecified change on the 12CE engine was the replacement of the steel vacuum pipe between front carburettor and distributor by a black plastic pipe.

An alternator replaced the dynamo in Australia from, it is thought, chassis number YGN4-1000 in April 1969. The alternator used an external control box, fitted to the bulkhead panel in the same position as the dynamo control box.

With the introduction of the 'facelift' HAN10/GAN5 versions, the voltage control box was changed to a three-bobbin RB340 unit, the black wiring loom covering changed from cotton to PVC, and the two-fuse system was replaced with a Lucas 7FJ fusebox containing four fuses and two spares stored vertically in the base. A separate in-line fuse was used for the new hazard warning circuit on North American versions from HAN9-72034 and GAN4-60441 (November 1967), and on home market Midgets from GAN5-138801 (August 1973).

The foot-operated dipswitch on home market 'facelift' cars was replaced by a combined dipswitch/flasher and indicator stalk on the steering column. Initially, this control also operated the twin horns which were standardised on the 'facelift' versions, but from HAN10-86303 and GAN5-89515 (June/July 1970) the horn control returned to the centre of the steering wheel for all markets. Also at this point, an in-line fuse was added behind the fusebox to protect the heater blower and wiper motor circuits. BMC's novelty feature, the blocked oil filter warning light, found no supporters with 'Leylandisation', so it was deleted from home market 'facelift' versions – the North American dashboard never had this feature.

Side marker lights incorporating the reflex units appeared on each wing on 'facelift' North American Midgets. Rear number plate illumination was catered for by a small lamp unit contained within the inner end of each quarter bumper, replacing the single unit previously found above the number plate. Rear light units now contained a flatter lens retained by two screws, the chromed Mazak retaining bezel being dispensed with. The lens was red over amber on North American cars, while home market cars continued with amber over red. Courtesy lights for the cockpit and boot were introduced from HAN10-86303 and GAN5-89515 (July 1970), operated by switches in each door pillar and the right-hand edge of the boot opening. The cockpit light, mounted centrally below the dashboard, also had a sliding switch on its chromed cover.

Combined ignition and steering locks started to be applied to the 'facelift' versions, most probably as an option as no clear change point appears in the factory parts list. However, it would seem that all cars after HAN10-86766 and GAN5-96273 (January 1971) had a steering lock. A chromed blanking piece covered the redundant ignition switch hole on the dashboard.

The 'round-arch' Midget saw few changes on the home market. A Lucas alternator replaced the dynamo from GAN5-128263 (December 1972) with the introduction of the 12V 588FH engine, the voltage control box disappearing from the bulkhead panel. A 16ACR alternator was used at first, but a 17ACR arrived at engine number 12V 588 FH 2551 (12V 671 ZL 9570 for North America); the drive pulley was then reduced in diameter at 12V 588 FH 3193 (12V 671 ZL 12846 for North America) to increase rotor speed. The hazard warning lights fitted from GAN5-138801 (August 1973) were operated by a switch on the right-hand end of the dashboard. With the introduction of the 12V 778 FH engine in September 1973, the battery became type A98 and the ignition coil type 11C12.

For the North American market, an alternator was fitted to all 'round-arch' models, while a seat belt warning system arrived soon after at GAN5-112276 (December 1971). A seat cushion sensor detected whether the seat was occupied, while a switch built into the seat belt buckle was operated when the belt was fastened. An erring driver or passenger would be warned by a light and a buzzer sound as soon as first or reverse gear was selected with the ignition turned on. The buzzer also sounded if the driver's door was opened with the ignition key still in the switch. This system became more sophisticated from GAN5-138801 (August 1973) with the addition of a starter motor isolating relay. With this 'sequential' system, the starter motor would not operate unless both occupants were wearing their seat belts and the gear lever was in neutral!

From GAN5-123751 (August 1972), all dashboard switches on North American cars (and those for West Germany and Sweden) were illuminated, and a rheostat control above the rev counter replaced the on/off panel light switch.

While the front side light/flasher light assemblies remained unchanged throughout the HAN9/10 and GAN4/5 period on home market cars, there were differences elsewhere. West German cars continued to use a flasher-only unit with an amber lens, the sidelight being part of the headlight. North American cars used a twin-filament bulb for side and flasher, early cars having a white lens until an amber lens arrived some time in 1967; the lens pattern changed from concentric circles to vertical ribs with the introduction of the 'facelift' GAN5 Midget.

Number plate lamps continued unchanged for all

The change from positive to negative earthing brought with it a new battery spill tray, the black tray being the later type.

Sprite MkIV & Midget MkIII

Tucked away under the wing, the negative earthed cars used a new round-bodied wiper motor, still single-speed on home market cars. The air filter is a non-standard but more efficient K&N type.

home market GAN5 Midgets, but new lamp units for North America were fitted to the rear body panel either side of the licence plate from GAN5-143355 (December 1973) in order to accommodate the new energy-absorbing overriders. Headlight main beam ratings were 50W for North America and 45W elsewhere, but all cars had 40W dipped beams.

Transmission

The spring-type clutch was now replaced with a smaller 6.5in (165mm) diameter diaphragm clutch. This was still hydraulically-operated, but now from its own separate master cylinder. A larger ¼in (6.3mm) diameter hydraulic pipe ran down to the main frame rail where a welded bracket provided support for the connection between this pipe and a flexible hose to the slave cylinder. A special carbon graphite release bearing was still used, operating against a thrust pad on the centre of the diaphragm fingers.

The gearbox remained unchanged on HAN9/GAN4 versions except that the gear lever was now finished in black, to match the surrounding cowl. From 12CC-DA-12411, the remote control housing was changed to incorporate a reversing light switch operated by the gear lever, twin automatically-operated reversing lights having been added in September 1967.

With the introduction in October 1969 of the facelifted HAN10/GAN5 versions, a new gear lever and surrounding gaiter were introduced. The new lever contained a rubber bush, insulating the upper part of the lever from the base. The previous painted metal cowling around the base of the lever was superseded by a new black Vynide gaiter covering most of the lever beneath the knob.

A new gear lever knob for the 'round-arch' Midget replaced the traditional BMC item. Soon after, from engine number 12V 587 ZL 7591, the remote control housing and remote control shaft on North American cars were changed to accommodate a gear lever position sensing switch for the seat belt warning light and buzzer. Although GAN5-112276 (December 1971) was the first car equipped with this warning system, the gearbox revision slightly preceded this at GAN5-112036.

From GAN5-146370 (March 1974), the gearbox tailhousing and mounting on North American cars were adapted to incorporate a movement-restraining device comprising two main parts. Longitudinal movement of the engine and gearbox was restrained by a tie bar fixed between the bottom bellhousing bracket and a bracket sandwiched between the base of the gearbox-mounting bracket and crossmember. A rubber buffer was provided at the rear of the tie bar to allow for a controlled amount of movement between engine and body and to absorb noise and vibration. To the rear of the gearbox mounting, a strap over the top of the tailhousing limited any vertical movement of the gearbox in relation to the body. The ends of the strap were bolted to the gearbox-mounting bracket, with an insulating strip between the strap and tailhousing.

Rear Axle

The rear axle, still slightly wider for steel wheels than wire wheels, continued unchanged except for stronger halfshafts being fitted for both steel and wire wheel applications. The steel wheel shafts could be recognised by the new part number BTA 806 in raised letters on the end face. As the new shaft was interchangeable with the earlier type, it is now quite common to find one fitted to an earlier axle.

In December 1968, from HAN9-77591 and GAN4-66226, the crown wheel and pinion were replaced to give a higher ratio of 3.9:1. Road speed in top gear rose to 16.5mph per 1000rpm.

Front Suspension

No changes were made until August 1972, when the coil springs were increased to a free length of 9.85in (250.2mm) at GAN5-123837. In August 1973, the optional anti-roll bar became standard on home market Midgets at GAN5-138801. Otherwise, the front suspension remained unaltered for the 1275cc-engined versions.

Rear Suspension

Apart from the leaf springs being uprated to 80lb sq/in with a free camber of 4.72in (119.9mm), no changes were made.

Steering

From 1 January 1968, North American safety regulations required cars for that market to be fitted with collapsible, energy-absorbing steering columns. This took effect from HAN9-72034 and GAN4-60441 (November 1967), but home market cars continued with the original column.

The collapsible column required a mounting flange in the footwell panel, to which the perforated outer column was secured by three nuts and screws. The top of the outer column was provided with a mounting flange to bolt up under the bulkhead panel with three bolts. The inner column now comprised two shafts, the upper shaft being partially of hollow section to engage the lower shaft, which would slide within the upper shaft in a heavy frontal impact. Should the driver strike the steering wheel, the outer perforated column would collapse and allow the wheel to move forward, absorbing the impact energy. The sliding portions of the upper and lower shaft were of 'D' section to ensure positive rotational engagement. The collapsible column was also fitted with a combined steering lock/ignition/starter control, similar to that fitted to some European market versions since 1961.

Again for North American cars only, the steering wheel was changed from HAN9-78826 and GAN4-67476 (December 1968). The rim, still of 15½in (393.5mm) diameter, became padded, and the three flat spokes each contained five holes.

A much simpler collapsible steering column, lacking the energy-absorbing outer column of the North American cars, appeared on home market cars from GAN5-105501 (October 1971). The upper and lower shafts of the inner column were held in position by a roll pin and four plastic detente plugs, all of which would snap off under impact and allow the column to contract as the lower shaft slid further into the upper shaft. A point of warning here: it is quite possible to break the pin and plugs when attempting to force the steering wheel from its splines and tapered seating face. A proper steering wheel puller should be used to avoid damage to the column.

The revised steering column introduced in October 1971 featured a steering lock as standard. A steering lock had also been fitted for all markets from January 1971, but, prior to this, a lock was fitted only for some export markets, or as an optional extra on home market cars from HAN10-85287 and GAN5-74886 (September 1969). Since the pattern of the steering lock and matching ignition starter switch varied according to market and year, care is required when obtaining a replacement to ensure that the lock and switch are compatible.

The steering rack, sourced from the Morris Minor, was revised from HAN9-72528 and GAN4-61167 (January/February 1968) with a modified pinion and seal, and replaced entirely by a Triumph-sourced unit in January 1972 (see 'Production Changes'). The new unit required revised mounting brackets which now carried an angled split line for the clamping caps, the new brackets being black-painted steel. New track rod ends were of the 'Sealed for Life' variety, losing the lubrication nipple. The steering ratio was very slightly lower, giving 2⅓ turns from lock to lock instead of the 2¼ turns of the earlier unit.

The introduction of the 'facelift' HAN10 and GAN5 versions saw the adoption on home market cars of a steering wheel similar to the North American one, with a padded rim and three flat spokes. The centre of the wheel did not contain a horn push button, the new boss featuring the usual MG or Austin motifs.

From HAN10-86302 and GAN 5-89514 (June 1970), however, the horn button returned to the steering wheel on right-hand drive cars. Although of similar pattern, this steering wheel contained a modified fitting to the hub boss. The three spokes were separated and secured to the boss by two screws per spoke, whereas the earlier wheel had spokes radiating from a central flange with three screws securing this to the boss. The upper spokes were also placed horizontally rather than at a slightly 'ten to two' position, providing a better view of the instruments.

The Sprite now lost the heraldic Austin motif and was to continue with a plain centre boss and horn push until the end of production in July 1971.

Tales of fingers becoming stuck in the holed spokes prompted a change to a similar steering wheel with slotted spokes from GAN5-123731 (July 1972). This may not have provided a complete answer: from GAN5-135882 (June 1973) the spokes became solid with just a shallow recess where each slot had been.

BRAKES

Changes for the HAN9/10 and GAN4/5 models were chiefly confined to the master cylinder and pedal box arrangement. The combined brake and clutch master cylinder was now replaced by a separate cylinder for each function, the brake cylinder having a larger reservoir.

During production, two types of single-circuit master cylinder were fitted. Externally, these were identical except for an identification mark in the

The combined brake and clutch master cylinder gave way to separate cylinders in a new pedal box for the 1275cc-engined 'Spridgets'. The blower unit was moved closer to the heater unit to allow space for this, but now left insufficient room for the air valve. Initially, this was re-located to the front of the ducting pipe behind the radiator grille, but from HAN9-71121 and GAN4-58854 a new blower intake housed the valve, as shown here. Copper hydraulic pipes are not original but are likely to last a lot longer.

SPRITE MKIV & MIDGET MKIII

The first generation of Rostyle wheel fitted as standard to the 'facelift' Spridgets from October 1969. Sometimes mistakenly believed to be alloy wheels, these are of all steel construction and 4½in rim width. The removeable centre cap is shown here with a non-original stick-on motif. Jacking points are subject to severe corrosion, so only sound cars should be jacked up as shown here.

Wire wheel hub nuts changed over the years. Up to the introduction of HAN8 and GAN3 models, 12tpi eared spinners can be recognised by the wording 'left (near) side' and 'right (off) side', with an arrow to show the direction to turn for removal. HAN8 and GAN3 onwards had 8tpi threads and the words '(near)' and '(off)' were deleted. The octagonal nuts replaced eared spinners for all markets by 1969. All types had left-hand threads for the right-hand side.

form of small concentric circles on the main body of one of the cylinders. This identified that the piston and seals of this cylinder were different, a fact which has to be considered when ordering a repair kit. A clear plastic extension tube was fitted to the filler cap thread for some European countries, the cap being screwed onto the top of this; fluid level was increased and could be checked without having to remove the cap. The operating pushrod was of fixed length, no adjustment for either the pushrod or its stop position being provided.

The foot pedal return spring, formerly attached between the pedal box and pushrod, was now relocated and replaced by a much longer spring between the pedal and an anchoring tab welded to the footwell top panel inside the cockpit. The pedal box had a removable top cover over the pushrod-to-pedal pivot. The box was held to the footwell top panel by eight ¼ UNF cross-head screws into welded nuts in the panel. The pedal box and pedals were painted black, while the brake and clutch master cylinders were nickel-plated.

The change to individual master cylinders also allowed the installation of a dual-circuit brake master cylinder from HAN9-72034 and GAN4-60441 (November 1967) for the North American market, the new master cylinder being fitted with a clear plastic reservoir to display the fluid level. The front and rear brakes were now served by separate hydraulic circuits. The brass connector union fitted to the inner front wing panel was replaced by a circuit pressure failure warning sensor; a dashboard warning light would glow if one of the two circuits failed. The brake lights were now operated by a switch mounted in the end of the pedal box and controlled directly by movement of the brake pedal, whereas single-circuit brake lights continued to be operated by the pressure switch in the brass union connector.

Both the single (European) and dual (North American) systems remained unchanged until early 1972, when the front flexible hoses were altered on the 'round-arch' Midget. Coinciding with the change of steering rack, this occurred intermittently from GAN5-114352 and on all Midgets from GAN5-114643. The front flexible hoses now connected to the steel pipes through the inner wing panel rather than by way of the bracket welded to the crossmember below the shock absorber arm. This bracket was now deleted, with the length of flexible and steel piping being revised to suit.

Friction materials for all 1275cc versions were Ferodo 2424F-GG front pads and AM8-FF rear shoes.

WHEELS & TYRES

The HAN9/GAN4 models continued with the standard steel disc wheel or optional wire wheel. Dunlop C41 crossply or optional SP68 radial tyres could be fitted to either type of wheel. During 1968, UK cars with wire wheels had their 'eared' spinners replaced by octagonal nuts, as already used on exported cars. All Australian-assembled Midgets had wire wheels.

The introduction of the 'facelift' HAN10/GAN5 versions saw the arrival of Rubery Owen Rostyle 4.5J×13 steel wheels. Finished in black with a silver rim, these wheels had a fake alloy look. A cap retained by a spring clip blanked the central hole in each wheel, and extended chrome-plated wheel nuts with domed heads were used. From around 1970, Michelin tyres began to be used in both crossply and radial form.

For North America only, and possibly only for 1970, the optional wire wheels were chrome-plated. All other markets had a painted wheel in silver-grey 'Aluminium' finish, or Silver Birch Metallic in Australia.

The 'round-arch' Midget received a more common Rostyle pattern as used on the MGB, Ford Capri, Ford Cortina 1600E and some Vauxhall models. Again, black was the predominant colour, with silver grey rims and 'spokes'. Size remained at 4.5J×13, and 145 radial tyres were now standard.

With either design of Rostyle wheel the spare was now stored outer face upwards on the boot floor, a longer securing bolt and a new washer clamping the wheel in place.

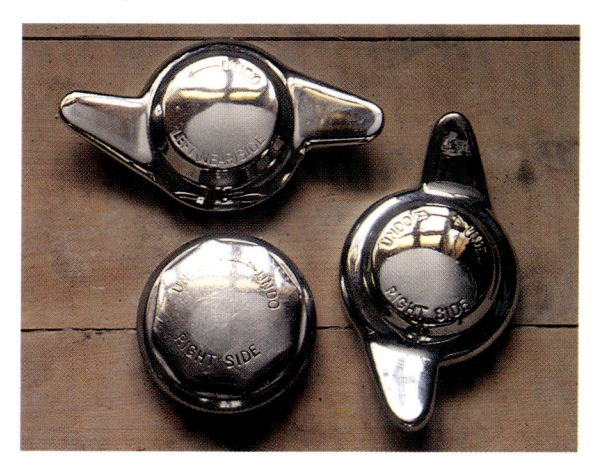

IDENTIFICATION, DATING & PRODUCTION

By now, the Midget was outselling the Sprite. The market seemed to favour MG, but the fact that BMC did not have to pay royalties on sales of Midgets may have prompted the manufacture of more Midgets than Sprites. The Sprite was withdrawn from all export markets during 1968-69.

The chassis plates for both cars were revised, the new plate having radiused ends. The space for the engine number, which was never stated, was replaced by the statement that seat belt anchorages complied with British standards number BS.AU.48-1965. The MG logo was dropped from the Midget chassis plate and replaced with the wording 'The MG Car Company Ltd, Abingdon-on-Thames, England'. This in turn was dropped, possibly in 1973, to be replaced with 'Austin Morris Group, British Leyland UK Ltd'. For home market production, the chassis number was also stamped into the driver's side front footwell just in front of the jacking point crossmember.

The body number prefixes and numbering sequence continued on from the previous versions until August 1970, when home market Sprite and Midget sequences were restarted at GBE 100101, both models now drawing numbers from the same sequence. Body numbers for exported GAN5 Midgets now had a modified prefix and sequence of numbers starting at GUN 150101 for North America or GSN 05000 for Sweden. From the beginning of 1967, cars leaving Abingdon carried the suffix G and the small batches assembled at Cowley M.

From 1967, North American cars were also now identified with the letter U to replace L immediately after the prefix. The letter U was later followed by a model year code letter, starting with A: thus a 1970 model year Midget for North America would become GAN5-U-A. The change from one model year to the next was identified by a new code letter accompanied by a small break in the GAN5 chassis number sequence, applied as follows:

A August 1969 to June 1970, 74886 to 88596
B June 1970 to May 1971, 89501 to 105146
C May 1971 to July 1972, 105501 to 123644
D July 1972 to August 1973, 123731 to 138753
E August 1973 to October 1974, 138801 to 153920

Also for North America, from January 1969, the chassis number had to be visible from outside the car, so a chassis plate was riveted to the top of the scuttle adjacent to the windscreen demister slot on the driver's side. The number was also stamped into the front right-hand shock absorber mounting, just behind the shock absorber body. A further plate, the main chassis plate, was riveted to the right-hand B-post door shut face and bore the statement: 'This vehicle conforms to all applicable Federal Motor Vehicle Safety Standards in effect on the date of manufacture shown above'. The month and year were stamped above this declaration with the chassis number below. The body number was still carried on a plate screwed to the left-hand door hinge pillar.

When a new version was approaching production, the pilot run was sometimes made before the previous model had finished its run. For example the very first 'round-arch' Midget was assembled in May 1971, some months before this version was announced and before the Sprite finished production in July 1971. There were, of course, no 'round-arch' Sprites other than the MkI! Also worth noting is the fact that the last five HAN9 Sprites, 85282 to 85286, were built as 'facelifted' HAN10 version, and were displayed on the Healey stand at Earls Court in 1969.

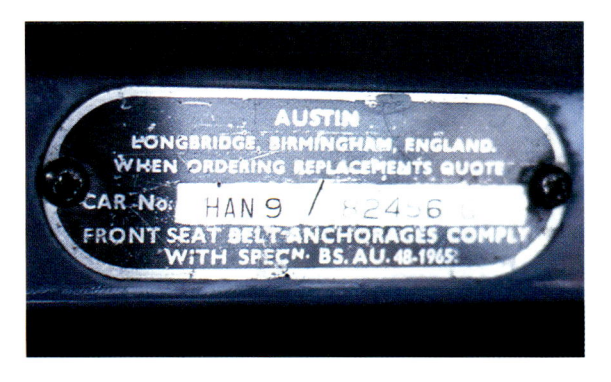

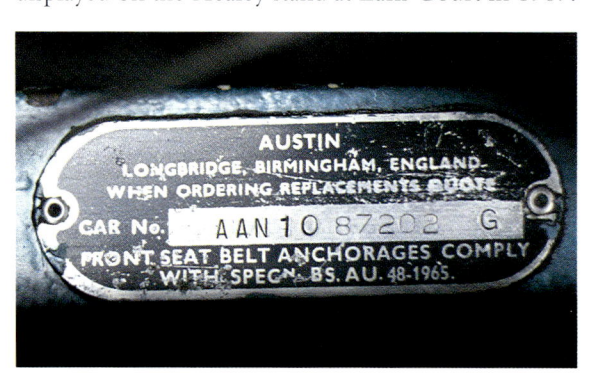

Three types of wheel found on 1275cc-engined cars. Right: the first generation of 4½in Rostyle wheel fitted as standard to all 'facelift' cars from October 1969. Top: second generation 4½in Rostyle which was standard for the 'round-arch' Midget, and never fitted to Sprites. Left: the optional 60-spoke wire wheel of 4in rim width.

A revised chassis number plate (far left) arrived for the 1275cc-engined 'Spridgets'. The space for the engine number was replaced with seat belt mounting information. A suffix letter also appeared at the start of 1967 to identify the assembly plant as G (Abingdon) or M (Cowley). The MG plate changed again around 1972 to delete the name of MG and replace it with 'Austin Morris Group, British Leyland UK Ltd'.

BL even saw fit to remove the H for Healey from the chassis number prefix of Sprites from January 1971. The Austin Sprite was identified by the modified prefix of AAN10.

SPRITE MKIV & MIDGET MKIII

PRODUCTION FIGURES

Year	RHD Home	RHD Export	LHD Export	LHD North America	LHD CKD	Total
Sprite MkIV (HAN9)						
1966	352	5	37	12	–	406
1967[1]	1270	83	339	5203	–	6895
1968	1005	38	164	5842	–	7049
1969	900	34	94	4979	–	6007
Total	3527	160	634	16036	–	**20357**
Sprite MkIV (HAN10)						
1969	129	–	–	–	–	129
1970	1282	–	–	–	–	1282
Total	1411	–	–	–	–	**1411**
Austin Sprite (AAN10)						
1971[2]	1022	–	–	–	–	1022
Midget MkIII (GAN4)						
1966	291	17	15	36	0	359
1967[3]	2853	169	470	4102	260	7854
1968	1790	96	235	4687	464	7272
1969	1726	71	139	4930	64	6930
Total	6660	353	859	13755	788	**22415**
Midget MkIII (GAN5, SQUARE-ARCH)						
1969	1196	154[4]		4605	80	6035
1970	3392	536[4]		10970	208	15106
1971	1657	180[4]		6458	108	8403
Total	6245	870[4]		22033	396	**29544**
Midget MkIII (GAN5, ROUND-ARCH)						
1971	430	83[4]		7553	–	8066
1972	4635	92[4]		11516	–	16243
1973	3737	0[4]		10311	–	14048
1974	1508	0[4]		8422	–	9930
Total	10310	175[4]		37802	–	**48287**

Notes to table
[1] The 1967 figure includes 489 Sprites assembled at Cowley between 24 January and 17 March. These Sprites carried a chassis number suffix M: eg, HAN9-65472M. From around this period, both Sprites and Midgets assembled at Abingdon now carried the suffix G. Cowley-assembled Sprites fell in the following batches: 65472-65496, 65641-65727, 65944-66092, 66195-66264, 66365-66434, 66585-66654, 66804-66820, and 67193; 190 chassis numbers allotted to Sprites for assembly at Cowley were unused.
[2] AAN10 Austin Sprites were produced between 27 January and 7 July.
[3] The 1967 figure includes 476 Midgets assembled at Cowley. Again, these cars carried an M suffix to the chassis numbers, which were: 53352-53376, 53473-53600, 53643-53697, 53754-53780, 53931-54001, 54151-54217, 54353-54454, and 54611.
[4] From 1969, RHD and LHD export models were bracketed together in production records, unless destined for North America.

ENGINE PREFIXES (1275cc)

Prefix	Number sequence	Application	Period
12CC	101 to 16348	All markets	Oct 66 to Jan 68
12CD	101 to 21184	North America emission control	Jan 68 to Aug 69
12CE	101 to 17639	Not North America	Jan 68 to Aug 71
12CJ	21201 to 43617	North America evaporative loss	Aug 69 to Jul 71
12V 586 FH	101 to 6217	Home market	Aug 71 to Dec 72
12V 588 FH	101 to 3240	Home market	Dec 72 to Aug 73
12V 778 FH	101 to 2457	Home market	Aug 73 to Oct 74
12V 587 ZL	101 to 15393	North America 8.0:1 compression	Aug 71 to Aug 72
12V 671 ZL	101 to 23958	North America 8.0:1 compression	Aug 72 to Oct 74

Notes to table
Not all numbers were used. With the 12CC, the first 1000 were issued but withdrawn because of a design fault. Only the 12CJ began at a number other than 101, appearing to be a continuation of the 12CD sequence. The sequence from 21201 to approximately 30000 may have been used for both 12CD and 12CJ engines.

111

PRODUCTION DATING

Date	Sprite	Midget	Notes
Oct 66	HAN9-64735	GAN4-52390	Production begins
Jan 67	HAN9-64965	GAN4-52798	First cars built in 1967
Mar 67	HAN9-67193	GAN4-54611	Production at Cowley ceases
Jan 68	HAN9-72163	GAN4-60871	First cars built in 1968
Jan 69	HAN9-79236	GAN4-67989	First cars built in 1969
Oct 69	HAN9-85286	GAN4-74885	Final HAN9 and GAN4 versions
Sep 69	HAN10-85287	GAN5-74886	First HAN10 and GAN5 versions
Jan 70	HAN10-85410	GAN5-81049	First cars built in 1970
Aug 70	HAN10-86190	GAN5-88596	Last 1970 model followed by break in numbers
Jun 70	HAN10-86301	GAN5-89501	Numbers restart, first 1971 model
Dec 70	HAN10-86802	–	Healey name dropped
Jan 71	–	GAN5-96853	First car built in 1971
Jan 71	AAN10-86803	–	First Austin Sprite produced on 27 Jan
Jul 71	–	GAN5-105146	Last 1971 model followed by break in numbers
May 71	–	GAN5-105501	Numbers restart, first 1972 model on 10 May
Jul 71	AAN10-87824	–	Last Austin Sprite produced on 6 Jul
Jan 72	–	GAN5-113617	First car built in 1972
Jun 72	–	GAN5-123644	Last 1972 model followed by break in numbers
Jul 72	–	GAN5-123731	Numbers restart, first 1973 model on 13 Jul
Jan 73	–	GAN5-129951	First car built in 1973
Aug 73	–	GAN5-138753	Last 1973 model followed by break in numbers
Aug 73	–	GAN5-138801	Numbers restart, first 1974 model on 23 Aug
Jan 74	–	GAN5-144039	First car built in 1974
Oct 74	–	GAN5-153920	Last 'round-arch' Midget produced on 17 Oct

No MkIV Sprites were assembled in Australia, the Midget MkIII instead taking over with the chassis prefix YGGN4, the first number being 501 built on 6 November 1967. The Abingdon chassis number was now omitted from the chassis plate, which was riveted to the bulkhead in the same position as on the Sprite. The 'Flying A' logo also disappeared at this time, but otherwise the information remained as before. The body number was stamped into the top right-hand corner of the passenger side upper footwell panel, usually hidden by the wiring loom.

At or around chassis number 1000, in May 1969, the prefix was shortened to YGN4, but the numbering sequence continued unbroken to the final alloted number of 1288 (or 1252, see later), manufactured in March 1970. Just prior to the prefix change, the practice of reverse stamping the chassis plates was adopted to give raised characters, the authorities believing that this would prevent illegal tampering!

From the middle of 1969, the chassis plate was omitted altogether, the number instead being stamped into the left-hand upper footwell panel near the battery. The number was produced using a conventional stamp rather than the reverse stamp. The deletion of the plate can be attributed to two factors. First, the plate was headed 'The British Motor Corporation (Australia) Pty Ltd', but by mid-1969 British Leyland had long since superseded BMC. Second, stocks of plates had expired and it was not felt worthwhile to produce a new interim plate, as

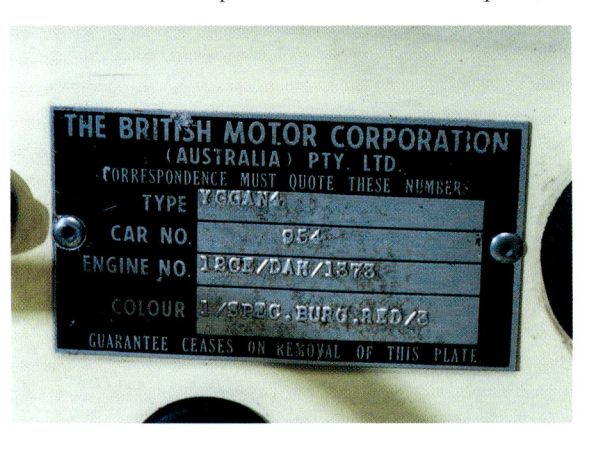

The chassis plate for the Australian-assembled YGGN4 Midget (far left), known locally as the MkI although in reality it was a MkIII! Champion Red paintwork is further identified by the prefix and suffix figures as being enamel paint by Balm paints.

Reverse stamping of the chassis number on Australian Midgets was introduced to prevent tampering with vehicle identification. In this instance (left), the original enamel Burgundy Red paint by Berger has been tampered with! This Midget has the unusual Australian YGGAN4 prefix.

Sprite MkIV & Midget MkIII

Conventional bulkhead-mounted chassis plates finished on Australian Midgets around June 1969, when stocks were exhausted. Without the chassis plate, the number was stamped directly into the upper footwell panel. A new type of identification plate was due for the beginning of 1970, so it was not felt worthwhile to produce a small number of old style plates. A sticker now served to give the paint specification.

As a further means to prevent tampering, BMC in Australia removed the engine number plate and stamped directly into the cast iron. This proved a more satisfactory practice than that used in Britain, missing engine number plates being common. Note the alternator bracket: Australia was first to install an alternator as standard on the Midget, from around April 1969 and YGN 1000. The bracket is different from that used for the later 12V engines. Australian engines were overpainted Longchamp Green.

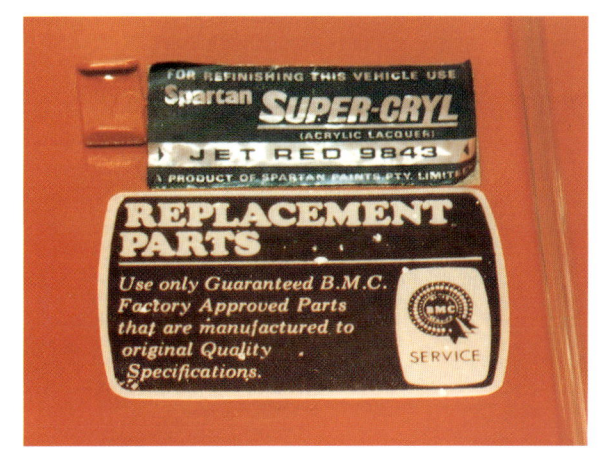

new compliance type plates were coming into force in 1970. Without a chassis plate, the body colour and paint manufacturer were displayed on a sticker placed just above the stamped number. A further sticker advised use only of BMC factory-approved replacement parts – shouldn't it have been BL parts?

The 'facelift' Midget, YGN5 with numbers starting at 501, became available in Australia from April 1970 and was fitted with the new compliance plate, again to the left-hand upper footwell panel. The plate stated that the vehicle was manufactured by British Leyland Motor Corporation of Australia, complied with Australian design rules, and was affixed with the approval of the Australian Motor Certification Board. Along with the chassis number, in raised characters again, the month/year of manufacture and seating capacity were also given. Paint colour and manufacturer continued on a separate sticker next to the plate. Shortly after the introduction of the compliance plate system, the plates were turned through 90° to read from the left-hand side of the car rather than the front, as all previous plates and markings had done.

An anomaly appears to exist in the stamping of one or more plates as YGGAN4 (with an extra A), this being repeated on the left-hand front shock absorber mounting platform. Whether this was the act of an inaccurate stamper or for some other meaningful reason is open to debate.

BMC (or BL Australia) removed the Coventry-applied aluminium engine number plates and stamped the same information directly into the cylinder block face, again to thwart the unscrupulous. Engines were over-painted Longchamp Green.

Although the rest of the world recognised the 1275cc Midget as the MkIII, the YGGN4 was often described in Australia as a MkI Midget and the 'facelift' YGN5 as a MkII, since all previous Australian-assembled cars had been Sprites. Assembly of CKD Midgets finished at YGN5-896 in December 1971, by which time sales were slow. Indeed, the final list price of AUS $2769 was marked down to AUS $2000 during 1972, the last example not being sold until September. No 'round-arch' GAN5 Midgets were assembled in Australia.

The total of 1184 Midgets assembled in Australia is thought to divide as 788 YGGN4/YGN and 396 YGN5, but there is an anomaly in the records. BMC Australia recorded the last of its 'MkI' Midgets as YGN4-1252, although the Abingdon figure of 788 CKD kits despatched (starting from chassis number 501) would make YGN4-1288 the last car. Should the Australian records be correct, then the YGGN4/YGN4 production figure would drop to 752 and the YGN5 figure would rise to 432, as *both* Abingdon and Australia agree on the total figure of 1184 Midgets for YGGN4/YGN4/YGN5.

Options, Extras & Accessories

The following items were available as factory or dealer fittings:

Heater Standard from HAN10-85287 and GAN5-74886.
Fresh air unit
Radio
Tonneau cover and rail Not available for Australian-assembled Midgets.
Laminated windscreen Standard for certain export markets.
Locking fuel filler cap
Cigar lighter Standard on North American Midgets from GAN5-105501.
Wing-mounted mirror Boomerang fitting. Driver's side mirror standard for North America but replaced with a door-mounted mirror from HAN9-72034 and GAN4-60441.
Wire wheels Standard on all Australian-assembled Midgets.
Anti-roll bar Standard on home market Midget from GAN5-138801 and on all Australian-assembled Midgets.
Headlamp flasher Standard on North American specification from HAN9-72041 and GAN4-60441, and for home market from HAN9-84674 and GAN4-73974.
Oil cooler Standard on all Australian-assembled Midgets.
Low 8.0:1 compression ratio Standard on North American Midgets from GAN5-105501.
Wheel trims
Whitewall tyres
Dunlop SP radial tyres 145-13 radial ply tyres became standard during 1971, but Michelin and Pirelli tyres were also being used by this time.

ORIGINAL SPRITE & MIDGET

Twin horns Standard from HAN10-85287 and GAN5-74886. Standard on all Australian Midgets
Luggage carrier
Hardtop
Rear compartment cushion Discontinued at the end of HAN9 and GAN4.
Front and rear chromed bumper guards Discontinued at the end of HAN9 and GAN4. North America only.
Supplementary tool kit In waterproof canvas bag, part number AKF 1596.

BMC Special Tuning became British Leyland Special Tuning when the amalgamation took effect in January 1968. Publications were C-AKD 5098 (BMC) and AKM 4168 (BL), both in loose-leaf form and re-issued at regular intervals. A huge variety of special tuning equipment was very quickly made available, categorised as follows (with part numbers):

Bodywork
Bonnet securing strap set, C-AJJ 3381; bonnet securing pin set, C-AJJ 4107; headlamp cowl kit, C-AJJ 3385; roll-over bar (rally), STR 0346.
Brakes
DS11 brake pad set, C-AHT 16; VG 95/1 lined brake shoes, C-8G 8997; VG 95/1 linings & rivets only, C-8G 8998.
Camshafts
Road, C-AEG 567; road/rally, C-AEA 800; rally, C-AEG 643; race, C-AEG 529; sprint, C-AEG 597; super sprint, C-AEG 595.
Carburettors
Twin 1½in SU HS4, C-AUD 709; installation kit 1½in SU, C-AJJ 4040; alloy flare pipe for 1½in SU, C-AHT 247; twin 1¾in SU, C-AUD 641; installation kit 1¾in SU, C-AJJ 4001; flare pipe for 1¾in SU, C-AHT 392; air cleaner 1½in SU, C-AHT 210; Weber 45 DCOE, C-AHT 143; installation kit and manifold for Weber, C-AJJ 3360.
Exhaust
Competition large bore manifold, C-AHT 11.
Clutch
Competition clutch cover assembly, C-AEG 546; competition clutch driven plate, C-AHT 383.
Distributor
Competition timing, C-27H 7766.
Cylinder head
Race large valve 16.4cc, C-AHT 221; race 16.4cc, C-AHT 222; polished 19cc, C-AHT 134; cylinder head joint, C-AHT 188; head washer set, C-AHT 288.
Pistons
Forged dished top (+.020in, +.040in), C-AJJ 3377; forged flat top (+.020in, +.040in), C-AJJ 3382; flat top (+.040in only), C-AEG 043043; forged flat top (73.5mm), STR 0310.
Valve gear
Tappet lightened, C-AEG 579; lengthened tappet adjusting screw, C-AEA 692; Duplex timing chain kit, C-AJJ 3325; valve guide set (Hidural), C-AJJ 4037; valve spring outer heavy, C-AEA 524; valve spring inner heavy, C-AEA 652; locating collar, C-AEA 654; inlet valve 1.401in (35.56mm) dia, C-AEG 544; inlet valve 1.464in (37.6mm) dia, C-AEG 55; inlet valve 1.307in (33.2mm) dia, C-AEG 569; valve rocker spacer, C-AEG 392.
Oil cooler and sump
Deep sump and oil pick-up kit, C-AJJ 3324; oil cooler kit, C-AJJ 3323; oil cooler cover, STR 0157; oil cooler (7 row), C-ARA 205; oil cooler (13 row), C-ARO 9809; oil cooler (16 row), C-ARO 9875.
Ignition
High tension kit, C-AJJ 4010; rubber plug connector, C-AHT 265; high tension cable, C-AHT 266; coil (HA12), C-AHT 269; Champion N57R plugs, C-27H 5982; N62R plugs, C-37H 2149; N60Y plugs, C-37H 2148; N64Y plugs, C-37H 4208; master cut-out switch, C-AHT 623.
Final drive and axle shafts
Limited slip differential, C-BTA 1226; axle shaft, heavy duty, disc wheels only, C-BTA 940; axle shaft, heavy duty, wire wheels only, C-BTA 939.
Gearbox
Straight-cut close-ratio gear set, C-AJJ 3319.
Suspension
Shock absorber competition front RH, C-AHA 6451; competition front LH, C-AHA 6452; adjustable rear RH, C-AHA 7906; adjustable rear LH, C-AHA 7907; lowered rear road springs (half-elliptic), C-AHA 8272; front suspension lowering kit to suit C-AHA 8272 springs, C-AJJ 3322; kit, anti-roll bar ⁹⁄₁₆in (14.3mm) dia, C-AJJ 3314; anti-roll bar, stiff ⅝in (15.9mm) dia, C-AHT 56; anti-roll bar, competition ¹¹⁄₁₆in (17.5mm) dia, C-AHT 57; installation kit for C-AHT 56 & C-AHT 57, C-AJJ 3356; rear bump stop kit, C-AJJ 4031.
Sundries
Blanking sleeve, thermostat by-pass, C-AJJ 4012; dual fuel pump kit, C-AJJ 4015; pulley for reduced speed, C-AEA 535; short fan belt, C-AEA 539.

COLOUR SCHEMES

The three accompanying tables list colour schemes throughout the Sprite MkIV and Midget MkIII period. Black is a continuing theme for much of the trim: the hood, tonneau, optional hardtop, dashboard and cockpit/door top trims were always black. Where piping in a contrasting colour remained available for a few of the HAN9/GAN4 colour schemes, it was applied to the seats and the door liners.

COLOUR SCHEMES (HAN9, GAN4)[1]

Body colour (code)	Seats/trim/carpet	Piping
Black (BK1)	Black	Grey
	Black	Black
	Red	Grey
Mineral Blue (BU9)[3]	Black	Black
Basilica Blue (BU11)[2]	Black	Grey
British Racing Green (GN29)	Black	Grey
	Black	Black[3]
Tartan Red (RD9)	Red[4]	Grey
	Black	Grey
	Black	Black
Old English White (WT3)[2]	Black	Grey
	Red	Grey
Snowberry White (WT4)[3]	Black	Grey
	Black	Black
Pale Primrose (YL12)[5]	Black	Grey
	Black	Black

Notes to table
[1] Hood, tonneau cover, hood cover and hardtop were available only in black for all Sprites and Midgets from October 1966.
[2] From November 1967, Mineral Blue and Snowberry White replaced Basilica Blue and Old English White.
[3] Sprite only.
[4] Midget only.
[5] Pale Primrose was new for the Midget, but had been available on the Sprite since late 1965.

COLOUR SCHEMES (HAN10, AAN10, GAN5)

Body colour (code)	Seats/trim/carpet	Period
Black (BK1)	Black	Throughout
British Racing Green (GN29)	Black	Until Jun 70
Glacier White (BLVC59)	Black or Autumn Leaf[1]	Oct 69 on
Bedouin (BLVC4)	Black or Autumn Leaf[1]	Jun 70 on
Pale Primrose (YL12)	Black	Until Jun 70
Blaze (BLVC16)	Black	Jun 70 on
Flame Red (BLVC61)	Black	Oct 69 on
New Racing Green (BLVC25)	Autumn Leaf	Jun 70 on
Blue Royale (BU38)	Black	Oct 69 to Jun 70
Midnight Blue (BLVC12)	Black	Jun 70 on
Bronze Yellow (BLVC15)	Black	Oct 69 on
Teal Blue (BLVC18)	Black	Jun 70 on

Note to table
[1] Autumn Leaf became available from HAN10-86301 and GAN5-89501.

PRODUCTION CHANGES

CHANGES BY CHASSIS NUMBER

HAN9-67193, GAN4-54611 (Jan-Mar 67)
Produced at Cowley with M chassis suffix.

HAN9-68228, GAN4-55035
Translucent extension tube fitted to the brake master cylinder filler for Benelux countries and France.

HAN9-69049, GAN4-56715 (Jun 67)
Interior locking levers fitted to both doors.

HAN9-69428, GAN4-56971 (Jun 67)
Front upper door seal changed.

HAN9-70268, GAN4-58112 (Sep 67)
Twin reversing lights included in rear body panel, operated by a gearbox-mounted switch.

COLOUR SCHEMES (ROUND-ARCH GAN5)

Body colour (code)	Seats/trim/carpet	Period
Black (BLVC122)	Navy	Until Jul 72
Bronze Yellow (BLVC15)	Navy	Until Aug 73
Glacier White (BLVC59)	Navy or Autumn Leaf[1]	Throughout
Flame Red (BLVC61)	Navy	Until Jul 72
Teal Blue (BLVC18)	Autumn Leaf or Ochre[2]	Throughout
Blaze (BLVC16)	Navy or Black[1]	Throughout
Green Mallard (BLVC22)	Autumn Leaf or Ochre[2]	May 71 to Aug 73
Aqua (BLVC60)	Navy	May 71 to Jul 72
Damask (BLVC99)	Navy or Black[1]	May 71 on
Harvest Gold (BLVC19)	Navy or Black[1]	May 71 on
Black Tulip (BLVC23)	Navy or Black[1] Ochre or Autumn Leaf[2]	Jul 72 on
Limeflower (BLVC20)	Navy	Jul 72 to Aug 73
Mirage (BLVC11)	Black	Aug 73 on
Bracken (BLVC93)	Autumn Leaf	Aug 73 on
Aconite (BLVC95)	Autumn Leaf	Aug 73 on
Citron (BLVC73)	Black	Aug 73 on
Tundra (BLVC94)	Autumn Leaf	Aug 73 on

Notes to table
[1] Second trim colour replaced the first in August 1973.
[2] Ochre was available only between GAN5-123731 (Aug 1972) and GAN5-138801 (Aug 1973).

HAN9-71121, GAN4-58854 (Sep 67)
Heater air intake flap valve moved from front right-hand splash panel to blower motor intake.

HAN9-71624, GAN4-59608 (Oct 67)
Starter solenoid changed, deleting starting button.

GAN4-59861 (Oct 67)
Pale Primrose paint becomes option on Midget.

HAN9-71911, GAN4-60100 (Nov 67)
Snowberry White paint becomes available.

HAN9-72034, GAN4-60441 (Nov 67)
First special North American specification versions with emission control engine. Padded facia with revised instrumentation and controls. Dual circuit brakes, collapsible energy-absorbing steering column, sun visors, electric screen washers and two-speed wipers. New quarter light assemblies, high-impact laminated windscreen, hazard warning lights. Rear-view mirror mounted to top screen frame rail.

HAN9-72041, GAN4-60460 (Nov 67)
Change to negative earth on RHD cars, necessitating a change of rev counter, wiper motor, fuel gauge with a voltage stabiliser. Temperature gauge changed to C-N-H graduation. Wiper arms and blades changed.

HAN9-72529 (Jan 68), GAN4-61166 (Feb 68)
Revised steering rack pinion shaft and seal. Revised direction indicator striker trip.

HAN9-72538, GAN4-61810 (Mar 68)
Door weather sealing rubber changed.

HAN9-72653, GAN4-60994 (Mar 68)
Last Old English White paintwork.

HAN9-72789, GAN4-60694 (Feb 68)
New single-piece exhaust system dispenses with sleeve joint to silencer, North America only.

HAN9-72964, GAN4-61674 (Feb 68)
Translucent extension tube fitted to the brake master cylinder filler (excluding Benelux countries, France, home market, and North America).

HAN9-73294, GAN4-61545 (Mar 68)
Additional breather pipe for fuel pump included to vent inside the boot.

HAN9-74462, GAN4-63075 (Jun 68)
Rear-view mirror changed on RHD cars, now contained in a grey plastic frame.

HAN9-74490, GAN4-62413 (Jun 68)
Revised brake pressure failure switch (North America).

HAN9-75703, GAN4-64475 (Jul 68)
Door quarterlight assemblies changed.

HAN9-76709, GAN4-65929 (Oct 68)
Temperature gauge changed to C-N-H reading (except North America).

HAN9-77591, GAN4-66226 (Dec 68)
Front sidelight units lowered in wings. 3.9:1 differential replaces 4.22:1. Windscreen wiper arms and blades changed. Hood and tonneau fitted with Velcro strips to secure to B-post. New seat covers, red seats and trim deleted. Tonneau support rails, rear cockpit moulding, hood frame and header rail now finished in black. Midget loses chromed bonnet trim moulding. On North American cars: bonnet hinges changed, head restraints fitted as standard, brake check light changed, three windscreen wipers.

HAN9-78731, GAN4-66224 (Dec 68)
Steering wheel and retaining nut revised.

HAN9-78826, GAN4-67476 (Dec 68)
North American cars: steering wheel changed, padded rim, three flat spokes each with five holes.
HAN9-80647, GAN4-69551 (Feb 69)
Revised brake pressure failure switch (North America).
HAN9-84674, GAN4-73974 (Aug 69)
Headlamp flasher now standard on RHD cars.
HAN9-85286, GAN4-74885 (Aug 69)
Last HAN9 and GAN4 prefixes.
HAN10-85287, GAN5-74886 (Sep 69)
Facelifted version introduced: 'Mk' designations unchanged but chassis prefix numbers advance by one.
HAN10-85307, GAN5-77803 (Oct 69)
Black windscreen fame discontinued and bright finish frame resumed.
GAN5-81742
Revised sun visors on cars to North America, Germany and Sweden.
HAN10-86301, GAN5-89501 (Oct 69)
Automatically engaging struts introduced for bonnet and boot lids.
HAN10-86303, GAN5-89515 (Jul 70)
Horn control returned to steering wheel centre boss. Revised steering wheel, now with horizontal upper spokes. Sprite now without a motif on the horn push. Door and boot courtesy lights fitted.
HAN10-86766, GAN5-96273 (Dec 70)
All home market cars given a steering lock.
AAN10-86803 (Jan 71)
First Austin-badged Sprite, prefix now AAN10.
GAN5-100179 (Mar 71)
Brake pedal changed.
GAN5-100745 (Mar 71)
Brake master cylinder push rod changed.
AAN10-87824 (Jul 71)
Last Austin Sprite.
GAN5-105501 (Oct 71)
First round rear wheelarch Midget. Larger fuel tank. New Navy trim replaces black, new door pull handles. New temperature gauge. Second-generation Rostyle wheels. Introduction of 12V type engines. Rocker switches on home market cars.
GAN5-112276 (Dec 71)
Hood storage cover changed. North American cars: changed hazard warning switch, new seat belt warning system with gearbox inhibitor switch and warning light set in radio console.
GAN5-112312 (Dec 71)
Navy head restraint changed.
GAN5-113000 (Dec 71)
Different indicator stalk control and lighting switch for North American cars.
GAN5-113617 (Jan 72)
Different reversing light units for North America.
GAN5-114588 (LHD), GAN5-114643 (RHD) (Feb 72)
Morris Minor type steering rack replaced by a Triumph unit with sealed-for-life track rod ends; steering arms and rack mountings changed to suit. Also intermittently from 114352 (LHD) and 114244 (RHD).
GAN5-114643 (Feb 72)
Front brake hoses changed to connect with steel pipes through the inner wing panels. Also intermittently from 114352 (Jan 72).
GAN5-115926 (Feb 72)
Steering lock changed on RHD cars.
GAN5-116170 (Mar 72)
Inertia reel seat belts for North American cars.
GAN5-118599 (Apr 72)
British Leyland badge on front wings changed from spire clip fitting to stick-on, now fitted to one wing only. Colour changed from blue/silver to silver/blue.
GAN5-119496 (May 72)
Different seat frames; new speedometer (LHD mph only).
GAN5-121650 (Jun 72)
Stainless steel quarterlight assemblies replace chromed units.
GAN5-123731 (Jul 72)
Steering wheel spokes slotted to prevent the irresistible urge to stick a finger in the previous holed spokes – and become stuck! Hood and tonneau cover bags changed. Autumn Leaf trim colour replaced with Ochre. Revised gear lever gaiter. North American cars: rev counter changed (RV1-1410/01AF), anti-burst door latches introduced, door units changed, sun visors changed.
GAN5-123751 (Aug 72)
Switchgear illuminated and instrument lighting now variable with a rheostat control, for cars to North America, Sweden and West Germany.
GAN5-123837 (Aug 72)
Longer front road springs.
GAN5-127605 (Dec 72)
Boot lid sealing rubber changed.
GAN5-128263 (Dec 72)
Introduction of 12V 588FH type engine. Alternator replaces dynamo on home market cars, rev counter changed (RVC 2415-01AR).
GAN5-129951 (Jan 73)
Changed reversing lights and rear lamp for North American cars.
GAN5-134125 (Mar 73)
Direction indicator striker trip on steering column changed.
GAN5-134186 (May 73)
Change to windscreen washer jets, washer bottle cap and pick-up pipe valve.
GAN5-135882 (Jun 73)
Steering wheel spokes now solid, not slotted.
GAN5-138801 (Aug 73)
All markets: Ochre trim dropped, Autumn Leaf reinstated. Black trim replaces Navy. Left- and right-hand rear exhaust mounting brackets changed. Front engine compartment splash shields revised. Seat belt mounting points on rear inner wheelarches repositioned slightly higher and forward. New front bumper mountings.
GAN5-138801 (Aug 73)
RHD cars: front anti-roll bar becomes standard, hazard warning lights fitted, anti-glare door mirrors fitted as standard to both doors.
GAN5-138801 (Aug 73)
North American cars: Revised steering column stalk controls and steering lock. Sequential seat belt warning system with starter motor isolation relay, warning buzzer changed. Revisions to bodyshell: rear panel, boot floor, rear inner wings, door units.

January 1971 to July 1971 saw the last 1022 Sprites badged simply as Austin Sprites, the long-running association with Healey having been terminated at the end of 1970.

Sprite MKIV & Midget MKIII

GAN5-139137 (Aug 73)
Flange fitting exhaust manifold to downpipe with single-piece exhaust system (home market).
GAN5-139307 (Sep 73)
Electric windscreen washer pump changed, LHD.
GAN5-139773 (Sep 73)
Twin-box, single-piece exhaust system now common for home and North America.
GAN5-141412 (Oct 73)
Speedometer now SN6142-06 BS on UK cars.
GAN5-143303 (Dec 73)
Combined courtesy light and ignition key warning buzzer door switch changed, driver's side only, on North American cars.
GAN5-143355 (Dec 73)
North American cars: front and rear licence plate support plates changed, licence plate lamps changed and fitted to rear panel, right-hand engine mounting changed, large energy-absorbing rubber buffer overriders.
GAN5-146370 (Mar 74)
Engine and gearbox restraining bar and strap introduced on North American cars.
GAN5-147531 (Apr 74)
Choke cable changed.

CHANGES BY ENGINE NUMBER

12CC-DA (all markets)

1000 (date unknown)
Engines to this number withdrawn due to a design fault, possibly crankshaft, oil pump and cylinder block weakness.
1386 to 1397 (Dec 66)
Nitrided EN40B crankshaft fitted.
1842 (Dec 66)
Nitrided EN40B crankshaft fitted, continuing to an unspecified point.
11639 (Jul 67)
Lucas 23 D4 distributor without vacuum advance unit replaced with a 25 D4 unit with vacuum advance.
12411 (Aug 67)
Gearbox remote control housing revised to include a reversing light switch.
16296 (Nov 67)
Oil pump changed.

12CD-DA-H (North America emission control)

356 (Dec 67)
Oil pump changed.
405 (Dec 67)
Improved rocker shaft.
1557 (Feb 68)
Revised main bearing cap dowels and bolts.
1745 (Feb 68)
Revised water pump with larger diameter inlet stub, lubrication plug deleted. Revised bottom hose.
2453 (Apr 68)
Tappet blocks revised.
8701 (Oct 68)
Vented carburettors replace closed circuit manifold valve.

12CE-DA (not North America)

874 (Feb 68)
Water pump revised in parallel with 12CD 1745.
899 (Feb 68)
Six-bladed plastic fan replaces steel fan.
959 (Mar 68)
Tappet blocks revised.
2942 (Apr 68)
Thermostat housing changed and adaptor fitted to cylinder head for temperature transmitter bulb.
3201 to 3300 and 3401 onwards (May 68)
Inlet manifold changed and vented carburettor bodies replace manifold control valve.
10309 (Oct 69)
Oil filter now disposable screw-fitting cartridge.

12CJ-DA-H (North America evaporative loss)

30847 (Jun 70)
Oil filter now a disposable screw-fitting cartridge.

12V 588 FH (home market)

2551 (Jul 73)
Lucas alternator changed from 16 ACR to 17 ACR.
3193 (Sep 73)
Smaller diameter alternator pulley, fan belt shortened by 1in to 32in.

12V 778 FH (home market)

393 (Nov 73)
Oil filter head changed.

12V 587 ZL (North America 8.0:1 compression)

7591 (Dec 71)
Gear lever, remote control housing and selector shaft revised to accommodate seat belt warning system inhibitor switch.

12V 671 ZL (North America 8.0:1 compression)

9570 (Jul 73)
Lucas alternator changed from 16ACR to 17 ACR.
12846 (Sep 73)
Smaller diameter alternator pulley.
13812 (Nov 73)
Oil filter head changed.

CHANGES BY BODY NUMBER

GBE-100650 (Home market), GUN-151300 (North America) (Sep 70)
New single heater/blower unit replaces individual units, modified mounting shelf and pipework to suit. Approximate corresponding chassis numbers were HAN10-86377, GAN5-91549 (home market) and GAN5-91407 (North America).

The adoption of automatically engaging struts for boot and bonnet from HAN10-86303 and GAN5-89501 (June 1970) made life much easier. The bonnet, however, was now inclined to sag to one side.

The round rear wheelarch lasted only three years (1971 to 1974) before a return to the original 'square' arch for the Midget 1500. Wheels were either second generation Rostyle or wire. Chromed wire wheels were offered for North American cars for a short time, but silver-painted wheels, as here, would otherwise be correct.

MIDGET 1500 (GAN6)

We have already seen that the North American market was the major recipient of Midgets. To continue selling in this important market, it was vital to meet increasingly stringent safety and emission control regulations. By 1974, BL had to make major revisions to the Midget if it were to continue.

The A-series engine was considered to be at the limit of economic development, but the need to use unleaded fuel and exhaust through a catalytic converter was on the horizon. To meet this, BL decided to equip the Midget with the Triumph Spitfire's 1493cc engine, which was far more suitable for meeting the new regulations without too much loss of performance. MG purists may have been aghast but the Midget would not have survived beyond 1974 without the Spitfire engine. A Marina-derived gearbox with synchromesh on all four forward gears accompanied this engine.

To meet low speed impact regulations, large energy-absorbing bumpers in black polyurethane replaced the slender chrome bumpers, earning the car the nickname of 'rubber-bumpered' Midget. The bodyshell required various modifications to suit, but perhaps the most surprising was the reversion to square wheelarches. In practice, it had been found that the round rear wheelarches did not provide such good resistance to rear-end impact damage.

The chassis number prefix became GAN6, although strictly speaking the A – for an A-series engine under 1200cc – should no longer have been applicable. The owner's handbook still referred to this Midget as a 'MkIII', but Midget 1500 or 'rubber-bumpered' Midget became common parlance.

BODY & CHASSIS

The addition of large front and rear energy-absorbing bumpers required a number of modifications to the bodyshell construction.

At the front, additional stiffener box sections ran along revised inner wing panels between the footwell and front panel assembly. The latter was now a part of the main structure and welded to the new stiffener

Bob Barker's very late Midget 1500 is one of the final 500 home market examples, all of which were finished in black.

MIDGET 1500

Gold side stripes and Rostyle wheels distinguish Michael Cohen's Midget 1500 as the one-off Jubilee version of 1975. The hood was unaltered when the Midget 1500 was introduced in October 1974.

rails and lower frame rails. The front valance and splash shields were also revised. The air ducting for the heater now ran along the left-hand side of the engine compartment into the air intake duct. The front suspension crossmember was revised to give increased ride height. Front wings, still bolted on, were revised with new apertures to provide clearance for the bumper-mounted indicator light units.

At the rear, additional stiffener rails were welded to the boot floor, running back from the boxed sections between inner wheelarch and floor. These additional stiffeners, front and rear, were the beam mountings to carry the front and rear bumper assemblies. A return to square-topped rear wheelarch was made to increase the strength of the rear structure, while the rear body panel was changed to suit the new bumper. Additional jacking points were included at the rear of each sill, the jacking tube being located in the inner sill and rear spring locating box.

Bodyshell units continued to differ between home and North American markets. Home market cars lacked the collapsible energy-absorbing steering column, burst-proof door catches, cantilever bonnet hinges, and the additional apertures for triple windscreen wipers and side marker lights.

Minor revisions to the engine compartment and transmission tunnel were required to accommodate the new engine and gearbox. Most noticeable was a hole in the bottom of the tunnel floor to allow access to the propshaft driving flange on the rear of the gearbox. It is possible that this hole was also included on some late GAN5 bodyshells as preparations were being made to press tools for the GAN6 bodyshell.

Access holes for clutch slave cylinder and gearbox oil filler for the A-series engine were not deleted from the GAN6 bodyshell until 1976, at GAN6-

The rear number plate was fitted below the bumper for the home market but above the bumper for North America. Later cars had black lamp covers either side of the number plate

Just as the original Sprite MkI earned itself an obvious nickname, so too did the Midget 1500. The 'rubber-bumpered' Midget was the longest-running version of all, and 72,289 examples were built.

119

There are no fuel pump breather pipes, but new rear bumper support beams and a re-aligned fuel filler hose are the recognition points of the GAN6 boot space.

175959 (North America) and GAN6-182084 (home). The rear inner wheelarch pressings were changed for the home market at GAN6-193666 (April 1977) when inertia reel seat belts became standard, North American Midgets already being so equipped.

BODY TRIM

Immediately apparent were the large black energy-absorbing bumpers front and rear. But despite the 'rubber-bumper' nickname, the bumpers were not made of rubber.

The front bumper comprised a black polycarbonate covering over a polyurethene foam filling. The rear was similarly constructed but with a black nylon covering, this material having better resistance against any fuel spillage from the filler cap sited just above the bumper on the right-hand side. A substantial steel armature was included in each bumper, making the complete units quite heavy. The front armature extended to a tubular frame around the air intake aperture, the armature losing its outer spring straps from GAN6-158716 (February 1975); the rear armature also changed in detail at the same time.

The new front bumper effectively replaced the need for a cosmetic radiator grille. However, to protect the radiator from stones, a simple black-painted mesh grille, of no visual value whatsoever, was placed behind the apertures in the bumper. The front indicators were incorporated into either end of the bumper, the sidelights being relocated into the headlight units. The bonnet no longer required the familiar aluminium trim piece to be riveted to its leading edge.

A small red MG badge, with silver lettering and octagon, was set in the centre of the air intake on the vertical spar. The boot badge, black with silver lettering and octagon, was the same as before. The BL motif on the front wing was deleted at an unspecified point during production. To mark 50 years of MG sports cars, some, if not all, Midgets produced in 1975 carried gold-coloured badges.

On 30 April 1975, a one-off Golden Jubilee

Gold badges were used in 1975 to mark 50 years of MG sports cars, even though the name MG can be traced back further than this. Not all 1975 Midgets, however, were so fitted.

The Golden Jubilee motif. Don't try to order this for your Midget, for only one was ever made.

The new front bumper also served as the radiator air intake, and there was a revised badge on the centre web. A rather agricultural cross-mesh grille behind the intake protected the radiator matrix from low-flying foreign objects.

Midget was produced in New Racing Green, with broad gold waistline stripes bearing the MG motif in body colour at the rear of each front wing. The Rostyle wheels were specially finished in gold and black, the steering wheel boss motif was gold, and a special commemorative plaque was fitted to the facia. This car was offered as a prize to the MG workforce, all of whom were given a raffle ticket. The lucky winner turned out to be a non-motorist who soon sold the car to a BL dealership. It survives today – as you can see in the photographs – and total mileage from new is less than 100.

In an attempt to revitalise falling sales in North America, a 'Midget Special' was offered from June 1976 for a limited period, and included 'free' optional extras as an incentive to potential purchasers. Standard features were chromed trims for the Rostyle wheels, a chromed luggage carrier and an AM/FM radio. More visibly, broad side stripes were added in the shape of a long, flat T, beginning from the rear of the front wheelarch and extending along the waistline to finish just above the rear side marker light. The vertical bar of the T sloped slightly forwards between the rear edge of the door and the front of the rear wheelarch. How many Midget Specials were sold is unclear, but most survivors will have lost their sidestripes if repainted.

The distinctive North American Midget 1500 Special, showing some of the standard 'extras' of luggage carrier, wheel trims and special side decals. A radio was also standard on the Special.

Tom Christensen's Midget displays a touch of personalisation along with a most distinctive licence number. Changes for North America included the repositioned licence plate holder with light units, sun visors, lamp lenses and optional dealer-supplied luggage rack, the latter being standard on the 1976 Midget Special. Black licence plate light covers replaced chromed covers from August 1976.

At the rear, the new bumper dictated a change to the number plate mounting and lighting arrangement. For home market Midgets, the rectangular number plate was fitted centrally *below* the bumper, with a lamp unit mounted to each end of the backing plate; the lamp units were the same as those used

ORIGINAL SPRITE & MIDGET

This thin type of head restraint was introduced in March 1977 and became a standard fitting for the home market.

Apart from its beige colour, the door trim on this 1979 car is identical to that used on the 'round-arch' Midget six years earlier.

Away from the dashboard, little differs in the North American cabin, but note the anti-burst door catch (below the main catch plate), the later slim head restraint and the different steering wheel.

Just how you might expect the interior to be after 80 miles from new!

singly on the pre-facelift 'Spridgets' until October 1969. North American Midgets used a licence plate mounted centrally *above* the bumper, again with a lamp unit on either side of the backing plate. At first, until GAN6-UG-179463 (June 1976), these were chromed lamp units carried over from the late North American 'round-arch' Midget. Revised light units were then fitted, with a chromed finish for a short time until GAN6-UH-182001 (August 1976), and black thereafter.

The rear lower wing quarters received black paintwork and a trim strip to match the familiar finish to the sills, which still carried the word Midget in individual chromed letters. An additional jacking point now appeared at the rear of each sill panel, a blanking grommet like the one used for the centre jacking point covering the access hole.

Revised anti-glare door mirrors were fitted from GAN6-175834 (March 1976) and again from

MIDGET 1500

The interior trim was virtually unchanged for the Midget 1500, seat coverings, trim panels and carpeting remaining unaltered. Beige trim, as seen here, replaced Autumn Leaf in October 1977.

Inertia reel seat belts did not become a standard fitting for the home market until April 1977.

GAN6-200001 (August 1977). There was a standard fitting for all markets, but the passenger side mirror was deleted at the second change point.

In the engine compartment, a different windscreen washer reservoir was introduced from GAN6-166301 (October 1975) on North American Midgets. The rigid plastic reservoir in the vee of the left-hand inner wing was replaced by a bag container hung from hooks beneath the wing drip rail to the outboard of the master cylinders. Home market Midgets continued with the rigid plastic reservoir, which was retained by a single strap located to welded tabs on the inner wing.

WEATHER EQUIPMENT

The hood, tonneau cover and hood stowage cover continued to be supplied only in black, with the tonneau cover now standard on home market cars. The tonneau cover was revised in February/March 1977 at GAN6-191678 (North America) and GAN6-192237 (home), with longer zips extending from the B-posts (behind the head restraint pockets) to finish level with the inside edge of the head restraint. The tonneau support rail was now discontinued from home market Midgets because head restraints were now standard.

The fibreglass hardtop, unchanged from the previous Midget, was still offered as an optional extra but gradually fell from favour and was discontinued at an unknown time.

INTERIOR TRIM

Despite the many other changes, the interior trim remained virtually unchanged for the duration of the 1500 Midget for both home and North American markets.

The pattern of seat covers, trim panels and carpeting remained the same as on previous models. Only the driver's footwell carpet was altered, a slot for the accelerator pedal no longer being required because a pendant pedal replaced the old floor-hinged organ pedal. Black trim continued throughout but Autumn Leaf was replaced by Beige from GAN6-200710 (October 1977).

Intertia reel seat belts replaced the static belts on home market cars from GAN6-193666 (April 1977), North American cars already having inertia reel belts. Belts were usually Kangol, the North American versions having a circuit switch for the warning system built into the driver's belt.

A revision to the black trim and seat coverings occurred in February/March 1977 at GAN6-191678 (North America) and GAN6-192237 (home). Accompanying this were new head restraints (without any wraparound shape) which now became standard on home market Midgets.

123

ORIGINAL SPRITE & MIDGET

The Golden Jubilee dashboard with unique treatment of the steering wheel (with a black finish for the spokes) and facia motif — but otherwise this is the style found on all home market Midgets up to August 1977. The optional centre console and radio are seen here.

DASHBOARD & INSTRUMENTS

Until significant revisions in August 1977, home market cars continued with the same layout of instruments and controls as before, except that the choke and air control knobs swapped positions. The rev counter (RVS 2415/01AR) contained an orange warning sector from 5500rpm to 6000rpm and a red sector beyond, but in practice even 5500rpm was not conducive to the longevity of the 1493cc engine. The speedometer was a Smiths SN6142/09S reading in 10mph increments to 120mph.

From GAN6-200001 (August 1977), new instruments, as shared by the Spitfire, were introduced. Most significantly, the oil pressure gauge was dropped and replaced by a green warning light in the bottom of the speedometer. The speedometer (SN 6211-125B) now swapped positions with the rev counter (RVC 2414/01F), which contained a red warning only from 6000rpm. The speedometer, reading to 120mph with dual markings in kph, contained the red ignition warning light and blue main beam warning light either side of the low oil pressure light. The trip recorder was re-set by a flexible drive cable, its control now being placed to the left of the instrument in the lower edge of the dashboard.

The dashboard lost its MG badge, but now a badge appeared on the radio blanking panel — unless an optional radio was fitted — below the speaker position in the centre console, the console being a standard fitting from GAN6-212001 (May 1978). A cigar lighter was fitted to the right of the speaker aperture at the same time, and above this position was a warning light for handbrake operation and brake circuit failure. The bonnet release knob, now a more convenient T-shaped handle in black plastic with a white symbol, moved to the lower left-hand transmission panel. The horn control moved from the steering wheel boss to the column stalk, which now had its various indicator, horn, headlamp dip and flash functions marked in white.

North American Midgets continued until the end of production with the same basic layout inherited

The later home market dashboard, seen on a 1979 car. Gone were the MG badge and oil pressure gauge, and the revised instruments were shared by the MGB and Spitfire. The speedometer and rev counter swopped sides, while the bonnet release handle, now T-shaped, was relocated. The centre console was now a standard fitting.

A late North American facia and console. Compared with earlier versions, only the seat belt warning light remains on the console, the EGRV and catalyst warning lights having been deleted for this non-Californian car. This four-spoke 'safety' steering wheel was fitted to all North American Midgets from August 1977.

124

MIDGET 1500

Familiar to Spitfire owners? The home market engine compartment with the 1493cc Triumph PE94/J engine, correct in every detail here.

The North American engine compartment as it was in 1976. The Exhaust Gas Recirculation Valve is mounted on top of the manifold, and the carburettor is a Stromberg CD4T with an automatic choke controlled by the temperature of the cooling water. The air inlet to the carburettor was also temperature-controlled by the blending valve to the left of the heater unit. To the left of this again is one of the two carbon absorption canisters used to contain petrol and crankcase vapours. Note the extra wide radiator used on North American 1500s. The air pump is hidden below the alternator.

from earlier versions. The choke control was deleted from GAN6-160160 (California, March 1975) and GAN6-166304 (elsewhere in North America, October 1975) when the automatic choke CD 4T carburettor was introduced. Also at this time, the brake circuit failure switch no longer contained a manual test function, this being incorporated into the starter circuit so that the light would glow when the starter was operated.

An Exhaust Gas Recirculator Valve (EGRV) service interval warning light was fitted to the centre console to the left of the radio blanking plate. On Californian versions, this was supplemented, on the right-hand side of the console below the cigar lighter, by a service replacement warning light for the catalytic converter. The EGRV warning light was deleted for Canada at GAN6-175273 (March 1976) and for Federal states excepting California at GAN-188000 (December 1976). California continued with it, but dispensed with the catalyst warning light at GAN6-212000 (May 1978).

As with home market cars, the instruments were similarly revised from GAN6-200001 (August 1977), with new markings and the removal of the all-important oil pressure gauge. An amber warning light now served this purpose, and was placed above the red ignition warning light on the right-hand edge of the instrument panel. Below these two lights was a blue light for high beam. The rev counter changed from RVC 1410/01AF to RVC 1414/00F, with a red

warning sector from 5500rpm. The speedometer, as in the UK, was calibrated to 120mph in 10mph increments. The temperature gauge was simply marked C and H, with small blue and red sectors. The fuel gauge, containing a small red sector near the E end of the scale, was marked 'unleaded fuel only' as all Midgets for North America were on a lead-free diet by this time.

From chassis number GAN6-229001 (October 1979), a plaque was added to the dashboard of home market cars to mark 50 years of the MG Midget from the Abingdon factory, the name Midget having been used until 1955 before being revived for the 'new' Midget in June 1961. Factory records show that these plaques were not all fitted at the factory, some having been posted to the dealer or customer at a later date.

ORIGINAL SPRITE & MIDGET

A late home market engine compartment, GAN6-212001 (May 1978) onwards, with dual circuit brake master cylinder, ducted air to the air filter box and without a BL motif on the rocker cover. This view is impossible to fault for originality.

ENGINE

The approach of even more stringent emission control regulations in North America prompted the change from the A-series engine to the type PE94/J Triumph Spitfire 1493cc engine. The cost of developing the A-series to accept unleaded fuel and 'detoxing' equipment was not felt worthwhile, especially since a more suitable engine was to hand within BL.

A brief look at the background of this engine may be of interest as it shared some surprising similarities with the A-series.

The Standard Motor Company launched its new small family saloon, the Standard Eight, in September 1953 as a head-on rival to the Austin A30 and Morris Minor. The Eight's engine was the same 803cc size as the A30's but its 26bhp at 4500rpm was 2bhp less. Next came the Standard Ten in October 1955 with a capacity increase to 948cc – the same as the enlarged A-series – and 33bhp at 4500rpm. The capacity coincidence ends here, for further enlargements were to 1147cc, 1296cc and 1493cc. However, the basic design of this engine remained recognisable throughout, like the A-series.

In addition to the historical parallels, there were also basic design similarities. The block and head were of cast iron with the rocker cover, sump and timing cover in pressed steel. Bore and stroke were 2.9in (73.7mm) × 3.44in (87.5mm). The crankshaft ran in three main bearings of 2.312in (58.72mm) with 1.875in (47.63mm) big end journals. Crankshaft end thrust was controlled by half washers in the upper rear main bearing housing only.

This last feature was to prove a weak link. Wear rate with half washers could often be rapid, especially on cars used extensively in traffic where prolonged use of the clutch would press the crank against the thrust washer. Ultimately, the rear washer would become so thin that it could fall out of the bearing housing, followed shortly by the front washer as the crank moved forward. Provided nothing went bang as these washers fell out, crank movement could then allow untold damage to occur – there are tales of connecting rods removing themselves from the engine.

Aluminium alloy pistons, of either solid or split skirt type, carried two compression rings and a slotted oil control ring. Gudgeon pin retention was by circlip and groove in each end of the pin bore. Connecting rod big end caps were split angularly and each was secured by two bushed dowels and bolts. Overbore size was +.020in (.508mm), pistons being available in 9:1 or 7.5:1 compression ratios. The latter was standard for North America, except for the 1976 model year when 9.0:1 was specified in all states except California. Cylinder liners were available to salvage oversize bores.

The crankshaft rear main bearing oil seal was a lip type carried in a separate housing bolted to the rear of the block. At the front of the crank, a keyed sprocket drove a single-row roller chain to the camshaft sprocket. The chain was tensioned by a spring-bladed slipper tensioner fitted to the pressed steel cover.

The camshaft normally ran in four directly-bored bearings in the left-hand side of the block, but later engines may contain bearing liners. A skew gear on the centre of the cam drove a near-vertical shaft to a Hobourn-Eaton eccentric lobe oil pump with a gauze strainer oil pick-up filter. The distributor engaged with a driving dog slot in the driven skew gear, the driven gear and distributor being located in a removable pedestal housing secured to the block by two studs and nuts. A pinch bolt passed through the pedestal to lock the distributor into position once the timing was set. Towards the rear of the camshaft, an additional eccentric drove the mechanical AC Delco fuel pump mounted to the left-hand side of the block and retained by two studs and nuts.

The oil pressure relief valve was located in the left-hand side of the block in a tapping angled towards the front. Above and forward of this was the seating face for the screw-on oil filter cartridge. On North American engines, an adaptor raised the position of the oil filter, in order to provide clearance for the air injection pump; these engines also had a twin-groove crankshaft pulley for a second vee belt to drive the air pump. A lip type oil seal in the timing cover prevented oil leaking past the pulley, while a cupped deflector washer was also fitted behind the pulley to throw oil rearwards. A groove in the pulley served as the timing mark, the crankcase angle being marked on a notched plate on the timing cover. Unlike the A-series engine which could only be read from beneath, this was more conveniently read from above.

The steel sump held 8 pints (4.5 litres) and was not fitted with any baffles. The sump-to-block flange was flat, the block walls extending downwards below the centre line of the crankshaft. The rear main bearing oil seal housing and a front main bearing sealing block ensured a continuous, flat sealing face for the sump.

As with the A-series, front and rear backplates served to locate the engine mountings at the front and starter motor and bellhousing at the rear. The starter was fitted to the left-hand side with two bolts.

The cast iron cylinder head, secured by ten studs and nuts, contained two vertically placed valves per cylinder. Valve head diameters were 1.380in (35mm) inlet and 1.170in (29.7mm) exhaust, and the valves ran in replaceable guides. Valve seat inserts and special exhaust valves were fitted on North American engines requiring unleaded fuel. Double valve springs were retained by split collets, but no valve stem oil seals were fitted. Each valve was served by its own port from the right-hand side of the cylinder head, the manifold face having round ports for the inlet and rectangular ports for the exhaust. Spark plugs were accommodated on the left-hand side between the pushrod apertures.

Bucket tappet blocks drove the pushrods to the overhead rockers with screw and locknut adjustment. The hollow, oil-fed rocker shaft was supported in four pedestals, each secured by a single stud and nut. Forged rocker arms were separated with spacing springs, except for the extreme end rockers which were located in extension lugs to the pedestals.

The pressed steel rocker cover, painted red on North American emission control engines, was retained by two studs and nuts. Later engines used blind-headed fasteners with a screwdriver slot in the domed head. Except on later engines, a small British Leyland sticker was applied to the silver/grey-painted cover adjacent to the vent pipe connection. The bayonet-fitting oil filler cap to the rear was non-venting, so a single stub pipe from the centre of the cover was the only provision for venting on this engine. On home markets Midgets, the vent pipe fed, via a Y piece, to both carburettors. On North American Midgets, the vent pipe was joined by a purge pipe from the carbon absorption canister before entering the single carburettor.

Cylinder head specification, which varied according to year and to local emission control regulations, is defined in the accompanying table.

Cooling water inlet and outlet was to the front face of the cylinder head, a combined water pump and thermostat housing being secured by three bolts to this face. The water pump impeller and bearing housing fitted into the front of this housing, while the thermostat went into the top with a water outlet casting above it. A cylinder block drain plug was provided on the right-hand side towards the rear.

The engine number was stamped in to a projection on the rear left-hand side of the cylinder block at head face level. All engines began with the prefix FP, while a suffix identified the application as follows: E, home market; UE, North America; UCE, California. Since this gives limited information, it can be seen that definitive identification of an engine has to be obtained from the cylinder head part number, which is visible near the manifold.

Incidentally, an FM-prefixed engine will have been taken from a Spitfire, a not uncommon occurrence following a rod-through-the-side experience. The engines are virtually identical, only the sump and exhaust manifold differing. Also, a YC prefix would indicate an ex-Triumph 1500TC or Dolomite 1500 engine. Engines were supplied to Abingdon from BL's Radford engine plant in Coventry.

The home market engine was rated at 65bhp at 5500rpm, with torque of 77lb ft at 3000rpm. The fact that this was some 6bhp down on the Spitfire is commonly blamed on an inferior design of exhaust manifold being used. It also seems a disappointing output compared with the 1275cc A-series, which in home market form was credited with 65bhp at 6000rpm and 72lb ft of torque at 3000rpm. But the standards of measuring horsepower changed over the years and may well have had a bearing on these figures, neither of these figures being the later and standardised DIN values of measured horsepower. Certainly, the Midget 1500 would appear to have had healthier horses than the 1275cc engine! The 1500 offered better acceleration and a higher top speed, despite its greater weight and imperfect gearing – but fuel consumption was markedly inferior. In the most strangulated North American specification, power output fell to 50bhp at 5000rpm with 67lb ft of torque at 2500rpm.

In conclusion, it does seem ironic that the A-series engine lives on today in fuel-injected and catalyst-equipped form in the Mini, while its 1974 replacement has long since disappeared.

COOLING SYSTEM

The cooling system for the home market contained 7.5 pints (4.3 litres), but North American versions used a larger radiator to increase capacity to 9.5 pints (11.4 US pints or 5.4 litres).

The radiator was contained within a black-painted steel cowling similar to that used on the earlier models. The expansion tank was fitted to the right-hand front corner of the engine compartment, and had a 15lb/sq in pressure cap. A seven-blade fan in orange plastic was fitted to home market cars, but a higher capacity fan in green was also offered for warmer climates. North American Midgets used a 13-blade viscous-coupled fan with a revised water pump assembly.

Water entered the pump housing from an outlet on the right-hand side of the radiator. Hot water leaving the cylinder head went into the same housing, but entered an upper passage to the thermostat, rated at 82°C (or 88°C for cooler climates). The housing contained two tappings on the right-hand side below the thermostat, the rear one for the water temperature sender unit and the front one for the hose connection to the water-heated inlet manifold. A T piece on the rear of the manifold directed water either back to the water pump housing via a hose and rigid pipe beneath the manifold, or to the heater matrix through the upper connection of the T. An on/off water tap was secured to the right-hand upper footwell panel to control the flow of water to the heater – the tap still had no provision for control from the cockpit.

Water leaving the heater fed into a connection in the rigid pipe leading forward to the water pump. North American Midgets with the 150 CD4T carburettor drew water from the rear of the inlet manifold to control the release of the automatic choke mechanism. Hot water leaving the engine passed through the thermostat housing, returning to an inlet stub on the left-hand side of the radiator. The top hose was secured to the top of the radiator cowling by a strap retained by the left-hand cowling bolt.

For home market Midgets, the expansion tank was connected directly to the radiator by a small-bore hose. This arrangement also applied to North American Midgets until GAN6-166303 (October 1975), but thereafter a modified radiator and thermostat housing was introduced. The hose from the top right-hand side of the radiator, now of larger bore, connected to a stub on the housing, from which a further stub pipe carried a small-bore pipe to a revised expansion tank. From here, a further pipe led to a plastic spill tank located in the vee of the inner wing, outboard of the air filter casing.

A Technical Service Bulletin was issued which appeared to call for this revision to be made retrospectively to earlier North American Midgets, but different part numbers were quoted for the expansion and spill tanks. The 'retro-fit' revision also included an adaptor containing a temperature-sensing valve unit to be fitted into the top hose. A vacuum line from the carburettor connected to this unit, which controlled a venting valve on the carburettor during warm-up.

A sticker on the top of the radiator cowling warned against removing the filler plug, located in the top of the thermostat housing, while the engine was hot. An anti-freeze advice sticker was also applied to the left-hand side of the cowling or to the bonnet slam panel.

The heater unit was similar to that used on the previous version but opposite handed, so that the fresh air ducting now passed along the left-hand side of the engine compartment to the front splash panel. The heater box continued to display a sticker warning that the unit was impossible to drain and that Bluecol anti-freeze must be used in cold conditions. An additional sticker warned of negative earthing of the battery, which was still placed immediately behind the heater unit.

CYLINDER HEAD SPECIFICATION

Part number	Application	Description
TKC 1155	Home Market, 1974-79	9.0:1, HS4
TKC 1409	North America (except California), 1974-76	7.5:1, CD4
TKC 2748	North American (except California), 1977	9.0:1, CD4, EPAI, ULF
TKC 3239	North America (except California), from 1977	7.5:1, CD4T, EMAI, ULF
TKC 1409	California, 1975-78	7.5:1, CD4T, EMAI, ULF
TKC 1410	California, 1978-79	7.5:1, CD4T, EPAI, ULF, GV

Key to table
Under the 'Description' heading, the compression ratio is followed by code letters as follows: **HS4**, two 1½in SU carburettors; **CD4**, single Zenith-Stromberg CD150 manual choke carburettor; **CD4T**, single Zenith-Stromberg CD150 auto choke carburettor; **EPAI**, exhaust port air injection; **EMAI**, exhaust manifold air injection; **ULF**, unleaded fuel and catalytic converter; **GV**, gulp valve in inlet manifold.

The heater and blower unit was virtually a left-handed version of that used in the later 1275cc-engined cars. The coil was moved to suit the 1493cc installation, which still kept all electrical ancillaries (fusebox excepted) on one side of the engine compartment. Dual-circuit brake piping now crossed the bulkhead panel on right-hand drive Midgets, the connections to front and rear circuits now being placed on the left-hand side of the engine compartment.

CARBURETTORS, FUEL SYSTEM & EMISSION CONTROL

Significant differences now existed between cars for the home and North American markets. Furthermore, California set a higher standard for emission control, requiring British Leyland to produce cars to suit three markets. Initially the 1493cc engine did not meet the new Californian regulations, causing a delay until the problem of using unleaded fuel had been answered.

Home market versions used two 1½in HS4 SU carburettors – enlarged versions of the HS2 used on previous models – on a water-heated alloy inlet manifold. Because the carburettors were now mounted on the right-hand side, the throttle cable was turned through a fairly small radius to operate on a mechanical linkage leading down to the interconnecting throttle spindle between the carburettors. Three return springs were still used but were relocated, the centre spring to a small bracket bolted to the exhaust manifold, the outer springs acting from above and anchored to a bracket bolted to the inlet manifold. The throttle cable and drop linkage were also mounted from this bracket.

A heat shield and carburettor heat insulator blocks were provided to reduce the heat transfer from the manifold to the carburettor. The choke cable connected to the rear carburettor with an interconnecting spindle to the front carburettor. Fuel metering needles were initially type ABT with blue-coded piston springs, but changed to ADN from engine number FP 59424.

Two paper element air filters were housed in a single casing made up of a backplate and cover. Two inlet stub pipes were included in raised portions on the cover, both stubs facing forwards, one above the other. The casing was painted silver-grey, or white on later examples. Four bolts passed through cover and backplate to secure the assembly to the carburettors with a support bracket from the centre of the backplate down to the exhaust manifold downpipe flange. To the top of the casing, a riveted plate bore the British Leyland motif and the part number of the casing, while a Unipart sticker was applied to the side of the casing towards the rear. Ducting pipes leading from the filter to the air intake grille were fitted to later examples.

The inlet manifold, with four inlet tracts to the cylinder head, contained both a balance pipe and a water heating passage, the latter fed by a short hose from the thermostat housing. A T piece on the rear of the manifold either returned the water to the water pump via a hose and return pipe or, when the heater tap was open, upwards from the T piece and through the heater system. The manifold provided horizontal mounting of the twin carburettors.

An AC Delco RA1 mechanical fuel pump mounted on the left-hand side of the cylinder block delivered fuel to the rear carburettor, an outlet being provided on the lid of the float chamber to pass fuel on to the front carburettor. The interconnecting pipe was secured to the air filter casing cover with a plastic clip, which had to be freed whenever the air filter was removed for servicing.

Fuel tank capacity remained at 7 imperial gallons (32 litres), but a longer filler neck was required because of the position of the rear bumper support rail which now ran along the boot floor.

In contrast to home market cars, North American Midgets initially used a single 1½in Zenith-Stromberg 150 CD4 carburettor on a new water-heated inlet manifold. This carburettor operated along the same principle as the SU, but contained three significant differences. First, the piston and metering needle were controlled by a rubber diaphragm rather than the piston being a precise sliding fit within the bore of a dashpot. Second, the float chamber was contained within the base of the carburettor body, the fuel now surrounding the main jet assembly with a double float either side of the jet. This was a neater installation than the SU remote type of float chamber, but mixture adjustment now required the use of a special tool inserted through the top of the carburettor. Third, the cold-start facility was by way of a separate fuel entry point on the side of the carburettor body. A rotating disc, controlled by the choke cable, determined the amount of additional fuel admitted for cold-start and warm-up. The fuel metering needle, similar to that of the SU carburettor, was type 44A running in a .100in (2.54mm) diameter main jet.

A new air filter casing, in white-painted steel (black on later versions), accompanied the single carburettor. The two halves of the casing were held together with four over-centre clips. The air intake was to the rear and connected by a short duct to an air temperature compensator valve unit placed on the right-hand upper footwell panel alongside the heater unit. A length of heat-resistant ducting led down to a hot air collecting tube on the exhaust manifold. The temperature compensator automatically controlled the temperature of the air being delivered to the air filter by blending hot air from around the exhaust manifold with cooler air drawn through the compensator's second air intake.

The alloy inlet manifold contained an intake tapping from the exhaust gas recirculator valve (EGRV). The valve unit, prior to GAN6-166304 (October 1975), was fitted to the front of the exhaust manifold downpipe, the gas recirculator pipe leading upwards to connect with the inlet manifold to the top of the carburettor-mounting flange. The purpose of the EGRV was to allow a controlled amount of exhaust gas to combine with the fuel/air mixture, in order to reduce nitrogen oxide exhaust emissions. The EGRV was controlled by a vacuum-sensing pipe from the carburettor, a further sensing pipe from the valve to the choke mechanism preventing the valve from operating before the choke was pushed in fully.

To the rear of the manifold, a second tapping provided the connection to the anti-run on valve (AROV). A vacuum-sensing pipe for the distributor advance unit also connected to the carburettor, both distributor and EGRV pipes having flame traps.

The carburettor contained two venting stub pipes. To the rear, the float chamber venting stub was connected by rubber hose to the carbon absorption canister placed behind the air temperature compensating valve. The front stub pipe provided the vacuum purge connection for the rocker cover and absorption canister, drawing in crankcase and fuel vapours to be burned in the combustion chambers. An in-line fuel filter was now fitted between the mechanical fuel pump and carburettor.

The fuel tank on North American Midgets was fitted with a capacity-limiting chamber which reduced total contents to 6.5 imperial gallons (29 litres). The tank vented via the vapour separator tank mounted to the right-hand side member in the boot, the vent pipe from this leading forward to the engine compartment and connecting to the carbon absorption canister. The AROV was mounted in front of the canister, which it supplied with clean air from its own downwards-facing intake pipe. The AROV would operate when the ignition was switched off, allowing air to pass directly into the inlet manifold and weakening the mixture to an unburnable state.

This system served North American Midgets to GAN6-UG-166303 (October 1975) but excluded Californian versions which became available from GAN6-UF-160160 (March 1975). Californian regulations dictated that only lead-free fuel be used, this causing delay in the availability of the 1500 Midget while all other North American cars were still able to use leaded fuel. The fuel tank for lead-free fuel had a special filler neck with a flap valve which would only accept the nozzle of a lead-free pump.

On all Californian cars and other North American Midgets from GAN6-UG-166304 (October 1975), the carburettor was changed to a Zenith-Stromberg type CD 4T, which was basically the same as before except for the addition of an automatic choke. The choke mechanism released as the temperature rose in the cooling system, water pipes being routed through the water jacket unit on the rear of the CD 4T carburettor. The fuel metering needle was now type 45Q (45K for California) with a D9 needle for the automatic choke. This also marked the point when lead-free fuel became the mandatory diet for all new North American cars. The EGRV choke air bleed was deleted at this time.

The exhaust port air injection system differed from the previous arrangement on the A-series engine. The air pump, mounted beneath the alternator and driven from an extra vee groove on the crankshaft pulley, delivered air to a diverter valve, which was controlled by a vacuum-sensing pipe from the inlet manifold. The connection was made with a T piece inserted between the AROV pipe and manifold. Manifold depression would allow air either to be dumped to the atmosphere by the diverter valve, or to pass through the check valve and onto the exhaust port air injectors in the cylinder head.

Further revisions occurred for the 1977 model year in North America. A revised air pump removed the need for the diverter valve and vacuum-sensing pipe, and air injection was now by way of a single delivery pipe and adaptor into the exhaust manifold. Along with this, the exhaust and inlet manifolds were changed to allow the EGRV to be repositioned from beneath the inlet manifold to above it. This allowed easier access to the EGRV, which required servicing at 4000-mile intervals.

This attention was critical enough for a service warning indicator, a counter driven by the speedometer cable, to be included. Mounted to the pedal box, this unit had an input and output connection for the speedometer cables, with an odometer display giving the mileage between service intervals. A warning light on the centre console would glow as the indicator unit reached the service mileage. Once the EGRV had been serviced, basically by decarbonising it, the indicator could be reset to zero using a special key. A double-function service interval counter was fitted for a while, the second function being to monitor exhaust catalyst life; this was primarily for California and not retained for all states.

A secondary carbon absorption canister was added to the system from GAN6-200001 (August 1977), the primary canister now moving to the front of the engine compartment just to the right of the radiator. The venting pipes from the fuel tank and float chamber, along with the purge pipe, were redirected accordingly. The primary canister drew clean air from the secondary canister, which assumed the position vacated by the single-canister system. Three air pipes from the top of the secondary canister drew clean air from the atmosphere, two of these pipes being fitted to a Y piece and served by a single larger-bore pipe. Except for California, the AROV was deleted on this system.

For the final year of production, Midgets for California were equipped with an inlet manifold gulp valve. As with the earlier A-series engine, this allowed a controlled amount of air from the air pump to be passed into the inlet manifold under conditions of high manifold depression. The valve was mounted to the rear of the inlet manifold and drew air from a revised exhaust port air injection system. A return to individual exhaust port injection only affected the Californian Midgets. This final revision called for a type 45R fuel metering needle and D9 choke needle.

Depending upon year and state regulations, an emission control information sticker was applied to the right-hand side of the bonnet slam panel.

The under-bonnet collection of pipes, pump, canisters, valves and flame traps gave the impression of an explosion in a spaghetti factory. Could this have been the reason why the rocker cover on a North American engine was painted red – the only way to find the engine?

Exhaust System

A cast iron exhaust manifold of four passages was produced with the centre pair cast together, all of these combining to form a single downpipe.

A tapping in the downpipe angled to the front right-hand side accommodated the exhaust gas recirculator valve (EGRV) for North American Midgets. On home market cars, this tapping was fitted with a blanking plug. From GAN6-166304 (October 1975), the manifold was changed to allow the EGRV to be fitted from above and the air injection pipe to deliver air directly into the manifold via a single tapping and adaptor in the downpipe.

The single downpipe, with a flanged fit to the manifold, passed across to the left-hand side of the car beneath the gearbox, to which it was secured with a saddle clamp. A flanged joint with a sealing olive connected the downpipe to the rear portion of the system, which used the same mounting points as the earlier twin-box system; one from the heel panel and two further mountings on the inlet and outlet of the transverse silencer. The two silencer boxes were new for the 1493cc engine.

From GAN6-160160 (March 1975), the Midget received a catalyst in the downpipe immediately below the manifold flange. Initially, the catalyst was for California only, but from GAN6-166304 (October 1975) all North American Midgets had one. The catalyst required replacement at 25000-mile intervals. Initially, a service indicator was included with the EGRV indicator mounted on the pedal box in the engine compartment, but the catalyst indicator was deleted at GAN6-212000 (May 1978). Various warnings concerned the care, operation and handling of the catalyst, for the ceramic core was very susceptible to damage due to its brittle nature and high operating temperature.

The single-piece rear system was changed twice on UK cars, at GAN6-172237 (January 1976) and GAN6-200001 (August 1977). The right-hand rear mounting bracket was replaced by a bobbin mounting from GAN6-169792 (November 1975).

Electrical Equipment & Lamps

The electrical equipment followed on from the previous version virtually unchanged apart from the revised location of the under-bonnet components.

The 1493cc engine carried all of its electrical equipment on the left-hand side, requiring a new wiring loom to connect up with the Lucas alternator, starter, distributor and coil. The coil was now located on right-hand drive cars on the left-hand upper footwell panel near the heater, but for left-hand drive it fitted onto the top plate of the pedal box assembly.

The starter motor was still type M35J, but the coil was now type 15C6 with a ballast resistor to give dual voltage for start-up and running. The distributor was type 45D4, but for North American versions from GAN6-166304 (October 1975) an electronic 45/DE4 with amplifier unit was used. Spark plugs were initially Champion N9Y, but changed to N12Y from engine number FP6458E. North American versions also used an anti-run on valve (AROV) located at the right-hand rear of the engine compartment with the carbon absorption canister. Except for California, the AROV was later deleted.

Staying with North American versions for the moment, the seat belt warning system was simplified from GAN6-159054 (February 1975), the gearbox inhibitor switch and seat cushion sensor being deleted. Unless silenced by the new circuit switch, a warning buzzer would sound for 8sec if the driver attempted to start the engine without first fastening the seat belt; the passenger seat belt was not fitted with a circuit switch. The console-mounted warning light also gave a visual reminder every time the ignition was switched on.

A matching warning light on the left-hand side of the console would glow when the EGRV service was due. This was supplemented, where applicable, with a warning light for catalytic converter replacement. It would appear that the service interval warning lights were either omitted for some states or phased out during production. Certainly, the EGRV warning light was deleted for Canada at GAN6-175273 (March 1976) and some, if not all, North American states from GAN6-188000 (December 1976). Only Californian versions may have continued with both warning circuits, although the catalyst warning light was certainly deleted at GAN6-212000 (May 1978).

From GAN6-200001 (August 1977) and for all markets, the oil pressure gauge was replaced by a warning light (on the dashboard for North American cars, in the speedometer elsewhere) wired to a pressure switch. The temperature gauge was now operated by an electric sender unit, instead of by a capillary tube sender bulb in the thermostat housing.

From GAN6-212001 (home market, May 1978) and GAN6-166304 (North America, October 1975), the handbrake was fitted with a switch to indicate that the lever should be released before driving off. How did we ever manage before? The warning light was on the centre console on home market Midgets and on the dashboard above the steering column on North American cars. The light also doubled as a circuit pressure failure warning for dual-circuit brakes, and as a bulb function check when the ignition key was turned to the start position. Also at this time, North American Midgets were fitted with a battery cut-off relay. Two-speed wipers, standard for America, did not appear on home market Midgets until GAN6-212001. They were controlled from a revised rocker switch on the dashboard; electric washers were provided for North America only.

All 1500 Midgets were equipped with amber front indicator lights, the sidelights now being included in the headlamp units. Sealed beam headlamps were rated as 60W/50W (main/dip) for the home market and 50W/40W for North America. Rear light units were unchanged from previous versions: home market cars used amber over red lenses and North American cars had red over amber. North American cars continued to use combined side marker and reflector units on front and rear wings. A factory-installed aerial became standardised from GAN6-212001 (May 1978) for home market cars and was fitted to the rear of the right-hand front wing.

TRANSMISSION

A single-plate diaphragm spring clutch of 7.25in (184mm) was fitted with a ball race release bearing. The hydraulic slave cylinder fitted into the left-hand side of the bellhousing. The actuating push rod operated onto the withdrawal lever from within the bellhousing.

The gearbox was based on that used in the Morris Marina and comprised three main cast iron casings – bellhousing, gear enclosure and tail housing. The tail housing also provided the gear lever location. A pressed steel plate was secured by eight bolts to the top of the gear casing to provide access to the gears. Synchromesh was now provided on all four forward gears, and the ratios were as follows: first, 3.41:1; second, 2.119:1; third, 1.433:1; fourth, 1.0:1; reverse, 3.753:1. Capacity was 1.5 pints (.85 litres) of EP 80W Hypoy oil; a filler/level plug was provided to the front left-hand side of the box with a drain plug in the base.

The tailshaft provided a flanged coupling to a suitably equipped propeller shaft. To gain access to the four nuts and bolts securing the coupling, a hole was provided in the bottom of the transmission tunnel immediately below the rear of the gearbox. To cater for the variance in the length of the shaft with rear axle movement, a sliding splined section was included just behind the front universal joint. The provision of a lubricating nipple on the shaft seemed a return to the old days, but this was to serve the sliding splines rather than a universal joint.

Early North American Midgets were fitted with a gear lever position inhibitor switch as a part of the seat belt warning system, but this was deleted from GAN6-159053 (February 1975).

REAR AXLE

The rear axles for either Rostyle or wire wheels remained unchanged until GAN6-182001 (August 1976), when the handbrake compensator bracket was deleted from the left-hand axle tube. The revision to the handbrake cable and linkage saw a smaller bracket welded to the rear of the left-hand axle tube. The bracket was drilled with two holes for the attachment of a rubber strap carrying the new handbrake cable mechanism.

The differential gear ratio was changed from 3.9:1 to 3.7:1 from GAN6-200001 (August 1977), a rather belated gesture to prevent the larger engine from

The Lockheed front disc brakes served from October 1962 until the end of production in December 1979. The most obvious change during this period was the repositioning of the flexible hose at its inboard end. The hose in this view connects with the rigid piping through the inner wing panel, earlier cars having had their connection via a bracket on the shock absorber mounting.

using higher revs than were either required or good for it! Road speed in top gear rose to 17.9mph per 1000rpm. The torque of the 1493cc engine was well able to take the higher gear ratio with corresponding benefits to fuel consumption and engine life, but without any noticeable effect on acceleration.

Front Suspension

The front suspension was carried over from the 'round-arch' Midget, but at GAN6-171478 (January 1976) the coil spring free length was increased from 9.85in (250.2mm) to 10.2in (259mm). This model's increased ride height was obtained by a revised front crossmember, the change to a longer spring further increasing ride height.

An anti-roll bar was still standard on all home market Midgets, but remained an option for North America until becoming standard at GAN6-203341 (December 1977). Meanwhile, the rubber bush bearing straps were revised on home market cars from GAN6-191113 (February 1977).

Rear Suspension

To increase ride height and support the extra weight, stronger six-leaf rear springs rated at 86lb/sq in were fitted to all 1500 Midgets. The spring free camber was increased from 4.72in (119.9mm) to 5.58in (141.7mm). The rear shock absorber drop link was revised from GAN6-157672 (January 1975), but otherwise the rear suspension remained unaltered to the end of production.

Steering

No changes occurred to the rack and pinion unit or steering arms, although revised steering columns, both still of the collapsible type, were introduced for both home (non-energy-absorbing) and North American (energy-absorbing) Midgets. The home market column was changed again at GAN6-170990 (December 1975), along with the toeboard rubber boot seal.

The steering wheel for both markets was carried over from the 'round-arch' Midget. However, the steering wheel boss, containing the MG badge and horn push, was changed at GAN6-157173 (January 1975), silver on black replacing red for the badge. For the North American market only from GAN6-200001 (August 1977), a new four-spoke safety wheel, with a large boss forming a web for the spokes, was introduced with a slightly revised steering column to suit. A large MG motif adorned the centre of the new wheel, which no longer contained the horn control. Home market versions also had the horn control moved to the indicator stalk, but the original steering wheel continued with a fixed boss. Shortly after this, from GAN6-200833 (October 1977), the combined steering lock and ignition switch assembly was revised.

Brakes

Home market Midgets continued initially with the existing single-circuit hydraulic system, but with minor revision to the pipework. A banjo union was now fitted to the pressure outlet at the rear of the master cylinder, allowing a two-way split of the pipework. The rear outlet fed the single pipe to the rear brakes, while the end of the banjo fed a single pipe running to a new three-way connector union, which was bolted to the right-hand inner wing panel and supplied individual pipes to each front disc brake.

The brake light switch was no longer contained in the hydraulic circuit. Instead, a mechanically-actuated switch was fitted into the end of the pedal box assembly for operation directly by the pedal and push rod.

From GAN6-200001 to GAN6-212000 (August 1977 to May 1978), a new master cylinder with a remote reservoir was used. The translucent plastic reservoir was mounted in a bracket from the left-hand side of the pedal box, a rubber hose connecting it to the pressure cylinder. Front and rear brake hoses were revised on the single-circuit system from GAN6-162565 (May 1975).

Dual-circuit braking did not arrive on the home market until GAN6-212001 (May 1978). Unlike the original North American system, this did not contain a manual circuit failure check switch. Instead, this function was included in the starter solenoid circuit, so that the brake warning light would glow whenever the starter was operated. The warning light also functioned as a handbrake reminder, a switch now being included in the handbrake assembly. The circuit fail-

Other than changes to spring rate and ride height, the basic rear suspension was unchanged from March 1964 to December 1979.

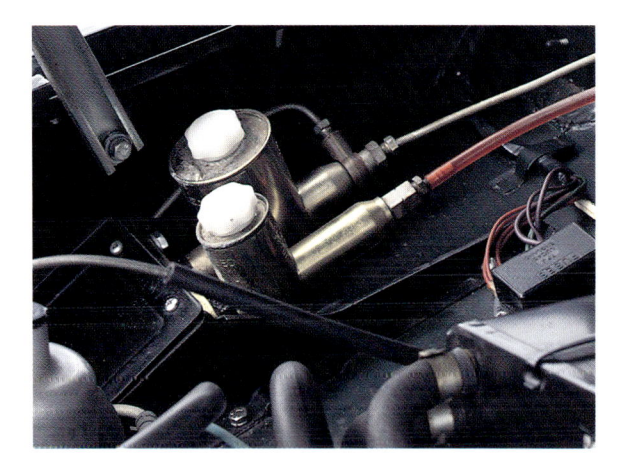

ure switch was fitted to the left-hand inner wing on all cars, but for right-hand drive it connected to the master cylinder with two pipes running across the bulkhead panel. North American Midgets had changed from the manual to automatic brake check system at GAN6-166304 (October 1975), the handbrake warning function also being added at the same time.

For all markets, the chromed handbrake lever was fitted with a plastic hand grip, a welcome change for owners in cold climates. No longer would bare fingers be frozen to an ice-cold handle first thing in the morning!

The handbrake cable and axle-mounting linkage were changed at GAN6-182001 (August 1976), when the pivoting compensator and reaction rods were deleted. A longer cable now acted directly on the left-hand rear brake, the outer cable being secured into the end of a frame-shaped bracket held to the axle by a rubber strap. The opposite end of the frame contained a cable extending to the right-hand rear brake. Brake balance was provided by the movement of the frame, the rubber strap allowing the measure of flexibility required but still supporting the frame and cables. Cable adjustment was now at the forward end where the outer cable entered the transmission tunnel. Two locknuts were employed to set the threaded portion of the outer cable against the body mounting bracket. A flattened P clip held the outer cable to the axle where it passed beneath the differential case.

Despite all these developments, the front discs and rear drums themselves remained unchanged for the Midget 1500.

WHEELS & TYRES

Rostyle 4.5J×13 wheels fitted with 145–13 radial tyres continued to be standard throughout production, but chromed rim trims became an optional extra for North America.

The wire wheel option also continued, but only until 1976 on the home market. Very few Midget 1500s appear to have been ordered with wire wheels for the home market, although they were still popular in North America.

IDENTIFICATION, DATING & PRODUCTION

On home market Midgets, the chassis number was carried on a plate on the left-hand side member on the front inner wheelarch. The plate was headed 'Austin Morris Group, British Leyland UK Ltd', wording which had been used since late 1972. The driver's side footwell continued to be stamped with the chassis number in front of the jacking point crossmember. For North America, the number appeared as before on a dated declaration plate on the driver's side B-post. Added to the declaration were the maximum permitted front and rear axle loadings, and a statement that the vehicle complied with the bumper standards in force at the time of manufacture. The scuttle-mounted chassis number plate continued as before.

For all markets, the body number appeared on a plate attached to the bonnet slam panel immediately behind the release latch. As the plate was fixed into position at the Swindon body manufacturing plant, it was sprayed over in body colour at the Cowley paint shop.

All chassis numbers were prefixed GAN6, starting at 154101 and finishing at 229526. The A of the prefix, as we saw earlier, was supposed to identify an engine of under 1200cc. Although the rule was slightly bent by the 1275cc A-series engine, the 1493cc Triumph engine should certainly have dislodged this prefix letter for the next capacity index, which would have been H. This, however, was already in use for the MGB, so to save any confusion it was felt simpler to continue with the A for the Midget 1500.

North American Midgets continued to use the additional U prefix plus the year code letter. Thus, the use of prefixes in relation to chassis numbers was as follows:

GAN6-UF-154101 to 166193 Nov 1974 to Aug 1975
GAN6-UG-166301 to 181663 Aug 1975 to Aug 1976

The early right-hand drive master cylinders, showing the different piping arrangement from the GAN5 Midget, the clutch pipe being nylon. Brakes were single-circuit on home market cars until GAN6-212000 (May 1978).

The unique finish of the Rostyle wheels on the one-off Golden Jubilee Midget.

Dual-circuit brakes arrived on home market Midgets in May 1978. This larger translucent reservoir replaced the remotely-mounted reservoir of the later single-circuit arrangement used from August 1977.

MIDGET 1500

Rostyle wheels continued unchanged as standard wear until the end of production. Wire wheels found few buyers on the home market but remained popular for North America.

GAN6-UH-182001 to 198804 Aug 1976 to Aug 1977
GAN6-UJ-200001 to 212393 Aug 1977 to Jun 1978
GAN6-UL-212394 to 229000 Jun 1978 to Oct 1979
GAN6-UM-229001 to 229526 Oct 1979 to Dec 1979

Note: M may not have been used on the last few North American Midgets. These may well have been regarded as 1979-only year models.

Body numbers began at GB 47T 040001P for the home market or GU 47T 004001P for North America. The identity of engine numbers is explained in the 'Engine' section.

OPTIONS, EXTRAS & ACCESSORIES

The only options listed for the home market were head restraints (standard from early 1977), wire wheels (until 1976) and a hardtop (which was discontinued at an undefined point). A larger range was offered for North America, as follows:

Solid-state AM or AM/FM radio[1]
Tonneau cover
Whitewall radial tyres

For home market Midgets, the chassis number plate continued to follow a similar style but was re-positioned due to the addition of front bumper strengthening beams running along the inner wings. Mention of MG on the plate had disappeared around 1973.

Wire wheels
Chromed wheel trim rims for Rostyle wheels[1]
Luggage carrier, cover and spyder[1]
Front anti-roll bar[2]
Rubber floor mats
Wooden gear lever knob
Door edge guards

[1] Standard on 1976 Midget Special plus special side decals.
[2] Standard from GAN6-203341 (December 1977).

PRODUCTION CHANGES

CHANGES BY CHASSIS NUMBER

GAN6-154101 (Sep 74)
First allocated chassis number.
GAN6-156798 (Jan 75)
Windscreen wiper wheelboxes changed, RHD and LHD.
GAN6-157173 (Jan 75)
Horn push button revised.
GAN6-157177 (Jan 75)
Roller friction head restraint.
GAN6-157672 (Jan 75)
Rear shock absorber drop links changed.
GAN6-158716 (Feb 75)
Front and rear bumper armatures changed.
GAN6-159054 (Feb 75)
Inertia reel seat belts include a micro-switch for the seat belt warning system and a time-delay buzzer, replacing seat belt warning sensor in seat cushion, starter relay and gearbox inhibitor switch (North America).
GAN6-160160 (Mar 75)
Introduction of Californian version: different fuel tank filler neck for unleaded fuel, catalytic converter with warning light, automatic choke CD4T carburettor.
GAN6-162211 (May 75)
Exhaust pipe saddle support bracket changed (except California).
GAN6-162565 (May 75)
Front and rear flexible brake hoses revised.
GAN6-166301 (Oct 75)
North American cars fitted with new bag-type windscreen washer reservoir.
GAN6-166304 (Oct 75)
Revised brake pressure failure sensor switch, radiator spill tank, and fuel tank filler neck for unleaded fuel. Vacuum connection between carburettor and thermostat sensor in bottom hose deleted. Handbrake lever receives switch and leads for warning lamp. Battery cut-off relay (North America). Catalyst exhaust and CD4T carburettor with automatic choke also on non-Californian US cars. New distributor type 45/DE4. EGRV repositioned to the top of the exhaust manifold.
GAN6-166305 (Oct 75)
Radiator and expansion tank changed (North America).
GAN6-167343 (Oct 75)
Clutch master cylinder and push rod changed (North America).

135

ORIGINAL SPRITE & MIDGET

PRODUCTION FIGURES

Year	RHD Home	LHD North America	Total
1974	591	1922	2513
1975	2531	11947	14478
1976	3465	13414	16879
1977	3774	10555	14329
1978	4574	9738	14312
1979	2912	6866	9778
Total	17847	54442	**72289**

GAN6-169644 (Nov 75)
Clutch master cylinder and push rod changed (not North America).

GAN6-169792 (Nov 75)
Changed rear exhaust mounting: right-hand bracket replaced with a rubber bobbin mounting.

GAN6-170990 (Dec 75)
RHD steering column and toeboard seal changed.

GAN6-171431 (Jan 76)
Front bumper and boot lid badges revised.

GAN6-171478 (Jan 76)
Front road coil spring free length increased from 9.85in (250.2mm) to 10.2in (259mm).

GAN6-172237 (Jan 76)
Exhaust system changed (UK).

GAN6-175273 (Mar 76)
Exhaust gas re-circulator valve warning deleted (Canada).

GAN6-175834 (Mar 76)
Revised anti-glare door mirrors.

GAN6-175959 (Mar 76)
Deletion of access holes in footwell panel and transmission tunnel for the A-series clutch slave cylinder bleed nipple and gearbox filler plug (North America).

GAN6-179463 (Jun 76)
Chrome number plate light units introduced (North America).

GAN6-182001 (Aug 76)
Fuel tank, handbrake cable and linkage, and rear axle casing changed (RHD). Black number plate light units introduced (North America and Japan).

GAN6-182084 (Aug 76)
Deletion of access holes in footwell panel and transmission tunnel for the A-series clutch slave cylinder bleed nipple and gearbox filler plug (RHD).

GAN6-183083 (Sep 76)
Steering column cowl changed.

GAN6-183592 (Sep 76)
Wiper blades become black (LHD).

GAN6-183740 (Sep 76)
Bonnet and rubber buffers changed.

GAN6-188001 (Dec 76)
Revised throttle cable (RHD and LHD). EGRV warning light deleted on non-Californian US cars.

GAN6-191113 (Feb 77)
Different front anti-roll bar bearing straps under frame rails (RHD).

GAN6-191159 (Feb 77)
Changed seat belts with micro switch (North America).

GAN6-191678 (Feb 77)
Different black door liners, seat covers, tonneau and head restraints (LHD).

GAN6-192236 (Mar 77)
Tonneau support rail deleted.

GAN6-192237 (Mar 77)
Different black door liners, seat covers, tonneau cover and head restraints (home market). Head restraints now standard on home market cars.

GAN6-193666 (Apr 77)
Different rear inner wheelarch pressings, inertia reel seat belts become standard for home market.

GAN6-195452 (May 77)
Reversing light units changed.

GAN6-198400 (Sep 77)
Seat frames revised.

GAN6-200001 (Aug 77)
All markets: 3.7:1 differential replaces 3.9:1, altered steering wheel boss motif, horn switch moved to indicator stalk, MG motif removed from dashboard, revised front bulkhead top shroud panel and doors (except North America), remote reservoir for brake master cylinder. UK only: revised exhaust pipe, different driver's door mirror and passenger door mirror deleted, revised instruments, oil pressure gauge deleted.

A body number plate was now included on the bonnet slam panel for all markets.

Notes to table
From the factory records, there is evidence to suggest that chassis numbers were not necessarily used in numerical order. For example, 200001 (a home market car) was built on 9 Aug 1977, yet 200004 (a North American car) had been built on 11 Jul 1977.

PRODUCTION DATING

Date	Chassis number	Notes
Nov 74	154101	First car built on 19 Nov
Jan 75	156670	First car built in 1975
Mar 75	160160	First Californian Midget 1500
Apr 75	162100	Jubilee Edition built on 30 Apr
Aug 75	166193	Last 1975 model followed by a break in numbers
Aug 75	166301	First 1976 model
Jan 76	171356	First car built in 1976
Jan 76	181663	Last 1976 model followed by a break in numbers
Aug 76	182001	First 1976 '½' model
Dec 76	187529	Last 1976 '½' model followed by a break in numbers
Dec 76	188001	First 1977 model built on 14 Dec
Jan 77	188820	First car built in 1977
Oct 77	198804	Last 1977 model followed by a break in numbers
Aug 77	200001	First 1978 model built on 24 Aug
Jan 78	204350	First car built in 1978
Jun 78	210870	Last 1978 model followed by a break in numbers
May 78	212001	First 1979 model built on 9 May
Jan 79	219817	First car built in 1979
Oct 79	229001	First of the final black batch built on 3 Oct
Dec 79	229500	Last of the black batch built on 6 Dec
Dec 79	229526	Last Midget produced on 7 Dec, finished in black and retained by manufacturer

MIDGET 1500

COLOUR SCHEMES (GAN6)[1]

Body colour (paint code)	Seats/trim/carpet	Period
Harvest Gold (BLVC19)	Black	Until Feb 76
Damask Red (BLVC99)	Black	Until Feb 77
Citron Yellow (BLVC73)	Black	Until Feb 76
Black (BLVC23)	Black, Autumn Leaf or Beige[2]	Throughout
Glacier White (BLVC59)	Black or Autumn Leaf	Until Aug 77
Bracken (BLVC93)	Black or Autumn Leaf	Until Feb 76
Tundra (BLVC94)	Autumn Leaf	Until Feb 76
Flamenco (BLVC133)	Black	Approx Feb 75 to Aug 77
Tahiti Blue (BLVC65)	Black or Autumn Leaf	Approx Feb 75 to Aug 77
Chartreuse Yellow (BLVC167)	Black	Approx Feb 76 to Aug 77
Sandglow (BLVC63)	Autumn Leaf	Approx Feb 76 to Aug 77
Brooklands Green (BLVC169)	Autumn Leaf or Beige[2]	Approx Feb 76 on
Carmine Red (BLVC209)	Black or Beige[2]	Feb 77 on
Leyland White (BLVC206)	Black	Aug 77 to Jun 78
Inca Yellow (BLVC207)	Black	Aug 77 on
Vermillion (BLVC118)	Black	Aug 77 on
Pageant Blue (BLVC224)	Beige[2]	Aug 77 on
Russet Brown (BLVC205)	Beige[2]	Aug 77 on
Porcelain White (BLVC243)[3]	Black	Jun 78 on

Notes to table
[1] Hood, tonneau cover, hood cover and hardtop were in black for all colour schemes.
[2] Beige (with chestnut carpet) superseded Autumn Leaf at GAN6-200710 (Oct 77).
[3] Porcelain White also known as Ermine White or Leyland White.

GAN6-200004 (Jul 77)
North America only: additional carbon absorption canister, different carburettor heat shield, different pedal box (except California), revised steering column, revised instruments, four-spoke steering wheel.

GAN6-200710 (Oct 77)
Beige replaces Autumn Leaf trim colour.

GAN6-200833 (Oct 77)
Changed ignition switch/steering lock (North America).

GAN6-201653 (Nov 77)
Changed windscreen frame seals, sun visor and pivots (North America).

GAN6-202111 (Nov 77)
Revised gear lever spring plunger (RHD).

GAN6-202329 (Nov 77)
Revised gear lever spring plunger (North America).

GAN6-203341 (Dec 77)
Front anti-roll bar standardised, with new front bearing straps (North America).

GAN6-206267 (Feb 78)
Fuel tank filler neck grommet in rear panel changed.

GAN6-210748 (May 78)
Inertia reel seat belts changed (RHD and LHD).

GAN6-212001 (May 78)
Introduction of dual circuit brakes with circuit failure and handbrake warning light, centre console, radio aerial, standard cigar lighter and two-speed wipers (RHD). Catalyst warning light deleted on Californian cars.

GAN6-212689 (Jun 78)
Revised clutch master cylinder and push rod (North America).

GAN6-212882 (Jun 78)
Revised clutch master cylinder and push rod (RHD).

GAN6-223510 (Jun 78)
Gearbox speedometer drive gear changed from 22 to 20 teeth (RHD).

GAN6-229001-229500 (Oct-Dec 79)
Last 500 home market cars painted Black, with Black or Beige trim, and 'Fifty-year' plaque fitted to facia panel.

GAN6-229526 (Dec 79)
Last Midget off the line on 7 December, black paintwork. Retained by British Leyland and now part of the British Motor Industry Heritage Trust collection.

CHANGES BY ENGINE NUMBER

FP1 (Sep 74)
First FP series engine.

FP 5332 (Feb 75)
Reverse gear operating lever changed.

FP 6458 (Mar 75)
Spark plugs changed from Champion N9Y to N12Y.

FP 59424 (Mar 78)
Carburettor needles changed from ABT to ADN (home market).

FP 79144 (Dec 79)
Last FP series engine.

The North American GAN6 chassis number plate only just fitted below the door anti-burst catch. Note the additional jacking point at the rear of the sill. The owner will no doubt have replaced the bung and the door draught excluder since this photograph was taken!

INNOCENTI

We have already seen that BMC used overseas subsidiaries to assemble cars from Completely Knocked Down (CKD) kits. There was generally an almost complete kit of parts, to which a small quantity of parts sourced or manufactured locally would be added to produce a complete Sprite or Midget.

An exception concerned the Italian market, where BMC already had an agreement with Innocenti in Milan for the assembly of the Austin A40 Farina – a British car with an Italian-designed body. For the Sprite, however, the Italian company used only the basic running gear in the form of engine, transmission, wheels, axles, suspension, brakes and steering. Everything else was designed and sourced locally.

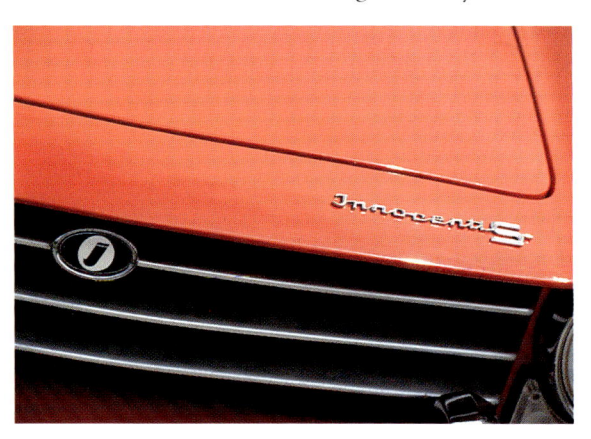

Ghia of Turin designed a completely new bodyshell based upon a similar floorpan to the Sprite. The bodyshell was produced by OSI using presses from Innocenti, the latter being by background a manufacturer of machine tools, hence the company's correct title of Societa Generale per l'Industria Metallurgical e Meccanica (General Metallurgical and Mechanical Industries).

Obviously, the company's full title would look rather laborious on a car badge – and would hardly impress the neighbours! So this 950 Spyder model has an Innocenti badge, using the company founder's name.

The Innocenti concept was to move upmarket from the cheap and cheerful MkI Sprite with a far more luxurious and stylish sports car which would also be much more practical to live with. As such, the 950 Spyder had everything the 'Frogeye' lacked: wide doors with exterior handles, winding windows, a lockable boot, interior light, cigarette lighter, ash tray, lockable glovebox and a heater as standard. It also had

Neil Roger's Innocenti S, a Sprite in Ghia-designed clothing. Around 17,500 of these were built in Milan between 1961 and 1970, but none was sold in Britain.

Badging at the front consisted of chrome-plated script giving the car's full title and the stylised 'I' motif on the grille.

INNOCENTI

No Abingdon-built Sprite or Midget had such a luxurious boot, with a floor covering and sound deadening applied to the inside of the lid. The fuel filler cap was neatly concealed by the lid when closed, but care was obviously required not to spill fuel into the boot when filling up.

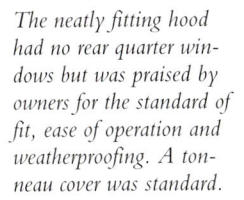

The neatly fitting hood had no rear quarter windows but was praised by owners for the standard of fit, ease of operation and weatherproofing. A tonneau cover was standard.

The Innocenti hardtop bore a resemblance to the BMC version offered for the Sprite MkII and Midget MkI, even if little else did.

Only the wheels and jacking point bung offer any clues to the Sprite connection. Ventilated wheels were discontinued in December 1962 at Abingdon, but are seen here on this 1966 Innocenti S.

Front wing badge recognises the roles of OSI as bodyshell manufacturer and Ghia as body designer.

a front-hinged bonnet which did not require practice to open and shut correctly, nor did it present the threat of being engulfed in the engine compartment should the side catches fail!

Overall, the 950 Spyder was slightly shorter but wider than the Sprite, although wheelbase and track were the same. The longer doors were achieved by placing the main bulkhead and scuttle further forward, giving a more spacious cockpit. However, the steering column remained the same length, placing the steering wheel just as close to the driver as in the Sprite. Inside, only the transmission tunnel, gear lever

139

An inviting cockpit and seats. It is extremely rare for the vinyl seat upholstery used by Innocenti to survive in this condition because it was not of the best quality.

and surround, and handbrake lever gave any clue to the Sprite connection.

Veglia instruments – speedometer, rev counter, fuel, oil and water gauges – were all neatly arranged in a binnacle in front of the driver. To the centre of the facia was the switchgear, and in front of the passenger was the lockable glovebox with a wide grab rail above. The top of the facia was sufficiently broad to allow the radio speaker to be set in the centre.

The windscreen was far more curved and swept back to meet fixed quarterlights at the front of the doors. The seats and hood were much improved over the Sprite items. The hood came in for particular praise for its quality, ease of operation and effective sealing – and the tonneau cover was standard.

Under the bonnet, the repositioned bulkhead left insufficient space for the heater and battery behind the engine, so the heater moved inside the cockpit and the battery sat in a box in the top of the passenger side footwell panel. The familiar 948cc A-series engine was equipped with a new air filter box and distributor. The radiator filler cap was positioned in

Veglia instruments provided the same information as for regular Sprites and Midgets.

A useful glove locker was yet another feature of the Innocenti's more lavish specification.

INNOCENTI

A familiar engine in unfamiliar surroundings. Innocenti used its own air filter casing, but otherwise the induction system was the same as on Abingdon-built cars. This Innocenti S should have a 1098cc 10CC-type engine (but with 10CB prefix for Innocenti), but the metal oil filler cap and breather pipe between rocker box and air filter suggest that this is a 948cc unit.

The Innocenti jack and tool kit. A maintenance manual was produced in English because these cars were sold in North America.

the centre of the header tank, there being sufficient clearance beneath the almost flat bonnet to avoid the expense of producing the extension filler neck used on the Sprite. Enclosed ducting fed air from the full-width radiator grille directly into the radiator matrix. To the right-hand side was the air inlet for the fresh air heater unit, and to the left of the ducting were the twin horns.

Externally, the 950 Spyder pre-dated the MkII Sprite in general shape but in a far more stylish manner. A hint of TR6 to the front, a touch of Alfa Romeo to the rear – it all came together to give an otherwise 'tin box' a very Italian character. Only the ventilated Sprite wheels looked familiar, but carried hub caps with the stylised I of Innocenti – and a really keen-eyed observer would have spotted the same jacking point rubber bungs.

Whereas the 'Frogeye' was simple in exterior adornment, the 950 Spyder was fully dressed. Wraparound chromed bumpers, sill and partial waistline trim mouldings, front wing repeater lamps, raised rear wing seam finishers, badges on bonnet, boot, front wings and grille badges, driver's side wing mirror and chromed exterior door handles all contributed to make this car not at all Sprite-like. Lighting units were also very stylish, with scalloped, triangular front indicator units beneath the headlamps and multi-function lamp units to the rear –

again pre-dating the MkII Sprite.

The 950 Spyder became available in 1961, just as its provider, the MkI Sprite, was nearing its end. Indeed, most of the early Innocentis were built up from what appeared to be leftovers from Abingdon. The last complete 'Frogeye' was AN5-49584 built in November 1960, most subsequent numbers up to the final AN5-50116 leaving as CKD kits for Milan. Although the 950 Spyder, perhaps rather oddly, carried AN5 chassis numbers, the engines had a new prefix (9CA-U-H) and sequence of numbers (101 to 750). Some 624 950 Spyders were built from AN5-sourced components before the HAN6 Sprite MkII took over.

141

ORIGINAL SPRITE & MIDGET

The forward position of the bulkhead left no room for the battery directly behind the engine. The electrical system was much better protected, having eight fuses compared with just two for the equivalent British Sprite. Note the central position of the radiator cap, extra space above the engine and radiator having allowed Innocenti to dispense with the extension neck. An under-bonnet light is mounted on the bulkhead adjacent to the heater duct.

A total of 4176 950 Spyders were built using HAN6 running gear and 9CB-prefixed engine numbers, Innocenti not drawing numbers from the Midget allocation at all. From late 1962 the HAN7 became the donor with disc front brakes and the 1098cc engine, designated 10CB for Innocenti. The car's title now changed to Innocenti S, which had a production run of 2880. Some improvements accompanied these updates: the steering wheel now had a semi-circular chromed horn ring to the lower half and the driving position was improved with, so it was claimed, more room for the taller driver.

The next update came with the HAN8 running gear in 1964, which meant the improved 1098cc engine (normally 10CC but 10CD for Innocenti) and semi-elliptic rear springs, Innocenti fitting telescopic rear shock absorbers rather than the lever arm type always used on Abingdon production. Innocenti now began its own series of chassis numbers, so numbers built were no longer recorded at Abingdon.

The dashboard layout was now revised with the oil, fuel and water gauges being placed in the centre of the painted facia panel. The raised binnacle in front of the driver disappeared, leaving the top of the new, flat facia following the radius of the scuttle. The top of the facia was still padded and covered, the radio speaker grille remaining in the centre. The ignition and starter moved from the right-hand side of the binnacle to the extreme left-hand edge of the new facia. Switchgear was revised with a centre console now being included. The passenger grab rail was deleted, but a padded roll was now provided right across the lower edge of the facia. Externally, large chromed overriders graced the bumpers.

The Innocenti S did not benefit from the 1275cc engine of the HAN9 Sprite. Local taxation law was unfavourable to the increase in engine size, so the 1098cc engine continued to be supplied especially for Innocenti.

The final update came in 1967 when a fixed coupé roof was added, the name becoming the Innocenti C Coupé. The radiator grille and sidelamps were also revised for this model. Production ceased in 1970 with around 17,500 Innocentis having been built. All were left-hand drive and sold chiefly in Italy, although there were some exports to North America and Switzerland.

Performance was very similar to the equivalent Abingdon Sprites: the Innocenti's extra weight took the edge off acceleration but slightly better aerodynamics allegedly allowed a higher top speed. Price was most certainly higher than the equivalent British models, so there was probably no space for the Innocenti on the home market between the 'Spridgets' and the MGB. In any case, BMC would probably not have welcomed it any more than the Triumph Spitfire!

The chassis number was stamped directly into the left-hand inner wing panel. Below it is the horn relay.

Paint chart shows a range of five colours, while English language brochure promotes the car's Italian character by showing three Innocenti S models high on the Monza banking.

142

BUYING & RESTORATION

With a production life of 21 years and 354,164 examples built, no-one could doubt the popularity of these cars, but many have succumbed over the years to leave considerably fewer than this figure today. Fortunately, in more recent times, the examples which survive are generally assured a much safer future in the hands of owners who are demonstrating a new-found regard for preserving and restoring their Sprites or Midgets.

In certain instances, this has come a little late for some of the earlier examples which were not so well regarded in the past. This is particularly true of the Sprite MkII (HAN6/7) and Midget MkI (GAN1/2) versions, which have become quite rare. MkI Sprites, on the other hand, have always enjoyed a special following with a healthy survival rate in the hands of owners who have generally been more inclined to preserve their cars than owners of late examples.

The 1275cc-engined cars are probably the most popular and practical versions to own and use on a regular basis. The later Midget 1500, although representing the largest single group with 72,289 examples built, was never so highly regarded because much of its re-engineering was intended simply to comply with regulations rather than to add driver appeal. The choice of which to buy must be a personal one, but the perils of buying a 'dud' must be recognised.

The cars which become available on the market can very loosely be thrown into three categories, regardless of model. First, there is the complete 'basket-case' for restoration. Expect to have to repair or replace everything and you won't be disappointed by a price which reflects condition. Missing parts or major accident damage are the main worries, but almost anything can be put right if you are looking for a challenging project. Second, a fully restored car will obviously command the highest price but this can often present good value for money long-term, providing that all the work has been carried out properly – which is sometimes difficult to ascertain if you are not completely familiar with the subject. Third, there is the 'in-between' jungle of semi-restored, repaired and generally cobbled-together cars. This is a very difficult area to buy into unless you are absolutely confident of your knowledge and fully prepared to meet whatever problems arise in a positive manner.

Thanks to the very basic design and engineering concept, repair and restoration work is very straightforward, and backed up by a wealth of experience obtained by the owners' clubs and trade establishments which have concentrated on these cars. The availability of spare parts could hardly be better and most components, even complete bodyshells, are now being re-manufactured. Indeed, it would be easier to list what is not available than what is.

This should not suggest, however, that a car with missing items can easily be put to rights. Some parts are peculiar to certain models, and it is essential to know what each model was fitted with and whether an absent component can be replaced, with an original, remanufactured or later specification substitute. In many instances, a part from a different model can be used to progress a restoration, but this will ultimately detract from originality and value.

The comprehensive catalogues available from leading parts suppliers will give the potential owner a detailed run-down of all the components which are currently available and, more important, the ones which are not. It should be remembered that many of the components now offered are not necessarily made by the original manufacturer or to the exact original specification. Although the possibilities of constructing an almost complete car from a catalogue may seem a reality, to do so would involve many compromises in original specification.

The introduction in 1991 by British Motor Heritage, a subsidiary of Rover, of complete new bodyshells suitable for Sprites HAN 9/10 and AAN10 and Midgets GAN4/5/6 has been highly significant in terms of the serious regard which owners now have for the long-term worth of their cars. Although the bodyshells are produced, loosely speaking, by the original manufacturer and on much of the original tooling, the new offering does not entirely conform to original specification for each of the types listed. What may be described as a 'universal compilation of specification' is used to suit both manufacturing expedience and compatibility between the different types. As such, both bonnet and boot lid are fitted with lugs for manual or automatic props or struts, no badge or trim strip holes are drilled, and the front radiator duct is produced with apertures for both left-hand (GAN6) and right-hand (pre-GAN6) heater ducting pipes. Rear panels are produced with hole patterns for quarter bumpers and reversing lights, leaving the owner of an early HAN9 or GAN4 version with more holes than required for the original features to be maintained. The Heritage bodyshell also utilises the later door latch mechanisms. It can be a convenient and cost-effective solution, but not the ideal in terms of originality.

Restoring a genuine bodyshell is perfectly practicable in many instances, but the cost can often exceed that of a Heritage bodyshell if this work is handed over to a professional. Owners of versions earlier than HAN9 or GAN4 or those wishing to retain as much of their original bodyshell as possible will have to face the task of a bodyshell restoration. Virtually all body panels and repair sections are available, but again an owner must be fully aware of the availability and suitability for his particular version before reaching for tin snips or chisel.

Thanks to the relatively simple design and construction of these cars, repairs to panel sections are often feasible when a complete and original section is

There's far more to choosing a car than deciding whether you want an Austin-Healey or an MG badge. Between 1958 and 1979 there were 14 distinct versions, four different engines and a staggering number of specification changes.

not available. Provided that a major accident has not distorted a bodyshell beyond redemption, a crucial point to check, then there is no reason why a restoration should not be physically possible. Undoubtedly, the main concern when tackling major rust repairs is to prevent distortion of the shell in the door gap area once all the corroded metal is removed. It is absolutely essential to support the shell correctly and to check frequently that the doors will still fit with parallel gap lines, using the doors rather as templates. Although the construction of the bodyshell is generally simple and straightforward, a good deal of care and patience is vital if the completed car is to feature good panel alignment and to retain all the strength that it was originally designed to have.

Mechanically, the design is again simple and straightforward, with no insurmountable problems even for the novice restorer. The A-series engines are generally very reliable and seldom fail without giving notice in the form of falling oil pressure and rising oil consumption. Only the 10CG-designated 1098cc engines should be singled out for having a slightly suspect crankshaft and bearings, but it usually takes a rather unmerciful driver to provoke problems.

The Triumph-based 1493cc engine used in the GAN6 has more justifiably acquired a reputation for crankshaft trouble, which can send a connecting rod through the block in extreme cases. The problem stems from the stretching of the original 803cc engine to 1493cc and its tendency to run rather hot in the confines of the Midget engine compartment, further exacerbated on earlier GAN6 models by a low 3.9:1 differential ratio. This engine does not like to be revved hard, its forte being a healthy output of torque. The crankshaft thrust washers can wear quite rapidly, allowing the crank to move sufficiently for the washers to fall out of their grooves and into the sump. A connecting rod can then catch an immovable part of the block and make a bid for freedom. An oil cooler is a sensible addition for this engine, and regular oil and filter changes are wise. It is surprising, therefore, that BL chose to omit the oil pressure gauge after chassis number GAN6-200001, leaving the owner with no visible warning of potential trouble.

The gearbox situation is rather the reverse. While GAN6 gearboxes seldom give trouble, the A-series 'boxes can suffer severe layshaft and laygear deterioration. This manifests itself as excessive noise in first and reverse gears. These were never quiet even when new, but the death rattle of an A-series 'box is quite unmistakable. The very early baulk ring synchromesh gearboxes always suffered a weak action on second gear, so jumping out of gear and gear lever chatter indicate a well-worn 'box.

Problems with the rear axle are mainly confined to the breakage of half shafts on earlier versions. The later specification half shafts, part number BTA 806, are a direct replacement on steel-wheeled cars and generally solve the problem unless a tuned engine is used. Differential gears and bearings last well, but oil leaks from the hubs are not uncommon – although easy to rectify. The front suspension can demonstrate an alarmingly high wear rate if not serviced or repaired properly; no self-respecting owner should be without a grease gun or forget to use it regularly.

Interior trim and seat coverings all suffer with time but virtually everything is replaceable and easy to fit. The main exception is the floor coverings of MkI/II Sprites and MkI Midgets, rubber matting of dubious quality which to date has not been remanufactured. Owners are obliged to use a carpet set from one of the later examples. To find one of these cars still with its original matting, complete and in good condition, is possibly one of the greatest treasures of all in Sprite and Midget folklore! A close second to this is the early style of steering wheel, which again has not so far been remanufactured. Any early car lacking its original steering wheel is quite likely to remain that way, unless the owner makes an extremely lucky discovery at an autojumble.

One golden rule when stripping a car for restoration is to label everything carefully as it is removed and to make notes of how items are fitted. Particular attention to pipe and cable runs along with the positioning of brackets and clips is essential. In addition, do not throw anything away no matter how bad or incorrect it may first appear. Everything is a point of reference even though it may not seem that way.

All too often owners throw away period accessories or modifications in ignorance of their significance. Some of these items can be extremely valuable and make the finished restoration stand out from all the other 'showroom' specification examples. Anyone discarding a Healey Motor Co disc brake conversion or supercharger kit just because Sprite MkIs did not leave the Abingdon factory with such items is not going to earn any Brownie points.

For unbiased help and advice, the potential purchaser or new owner should join one of the owners clubs appropriate to Sprites and Midgets. One can do little better than to talk to owners who maintain, restore, service and buy spare parts, and, most important, who enjoy their cars as they were meant to be enjoyed. The wealth of experience pooled in such clubs really can benefit anyone new to the cars.

The modest subscription fees can be likened to a small investment against buying the wrong car or doing the wrong things with it once you have it. Added to this, some clubs have discount facilities with parts suppliers and special insurance schemes which can more than save the cost of annual membership. A list of clubs is given, but the names and addresses of contacts, correct at the time of publication, may have changed by the time you read this. However, listings of specialist car clubs appear in some of the monthly classic car magazines.

CLUBS

Midget & Sprite Club
Terry Horler,
63 Littledean, Yate,
Bristol, BS17 4UQ
Tel: 0454 881770

MG Owners Club
Octagon House,
Swavesey,
Cambridge,
CB4 5QZ, UK
Tel: 0954 31125

MG Car Club
PO Box 251,
Abingdon, Oxon,
OX14 3FA, UK
Tel: 0235 555552

Emerald Necklace MG Register
Bob Zimmerman,
811 South Depeyster St, Kent, Ohio 44240,
USA

Austin-Healey Club
4 Saxby Street,
Leicester, LE2 0ND

Austin-Healey Sprite Drivers Club Inc
PO Box 248, Box Hill, Victoria 3128,
Australia

The Midget outsold the Sprite from 1964 and continued alone after July 1971 when the Sprite finished.